Pechan · de Vries
Genes on the Menu

Paul Pechan · Gert E. de Vries

Genes on the Menu

Facts for Knowledge-Based Decisions

With 6 color figures and 20 tables

 Springer

Dr. Paul Michael Pechan
Ludwig Maximilians University Munich
Communication Science Chair
Oettingenstr. 67
80538 Munich
Germany
e-mail: ppechan@compuserve.com

Dr. Gert E. de Vries
ProBio Partners
Meerweg 6
9625 PJ, Overschild
The Netherlands
e-mail: pbp@bart.nl

Library of Congress Control Number: 2004109051

ISBN 3-540-20178-5 Springer Berlin Heidelberg New York

Springer is a part of Springer Science+Business Media
springeronline.com
© Springer-Verlag Berlin Heidelberg 2005
Printed in Germany

Cover-design: design & production, Heidelberg
Typesetting: Fotosatz-Service Köhler GmbH, Würzburg

Printed on acid-free paper 52 / 3141 xv – 5 4 3 2 1 0

Foreword

When I received my university degree in plant breeding many years ago I was very excited about the new possibilities of developing better plant varieties. In the 1970s, tissue culture and micropropagation techniques became a welcomed addition to the breeder's repertoire. For example, protoplast fusion and microspore culture opened new ways of mixing plant genomes and the speeding up of plant breeding programs. Microbiologists had come up with a method of splicing genes predictably using restriction enzymes. We were happy to have these tools but we could not fully predict how empowering these tools would become to permit us to increase our fundamental knowledge of plants. The knowledge allowed us, in turn, to make much better decisions about which plants to choose and cross with each other in order to introduce useful agronomic traits into new plant varieties. At first, we were just glad to have these useful techniques: suddenly new and elegant possibilities appeared making it possible to improve crop yields and protect plants from various pests. What was started in 1970s became officially labeled, in the 1980s, as genetic engineering. A specific area of genetic engineering dealt with the genetic modification of organisms (GMOs). Very few people had any idea that this technology and its application in agriculture would, a few years later, become a focus of passionate public debate.

As scientists, our primary goal was to explore the unknown, learn from nature and find new ways to use this knowledge for the betterment of our society. We would report on these findings in scientific journals. We were not interested in politics and we were ill-prepared to be suddenly in the limelight, having to justify to a non-scientific public what we do and why. However, this we had to learn as the public concerns about the use and possible abuse of genetic engineering became more visible and vocal. A number of well-organized non-governmental organizations led the fight against the food and non-food use of GMOs. They voiced a number of concerns: some legitimate, some less so. Industry, in their initial insensitivity to public opinion, made the discussed issues even more emotionally charged – the fight against big companies wanting to control trade in agricultural products globally. Factual and ethical concerns became intermixed with political and ideological agendas and attitudes. What boiled to the top were not necessarily the important or the right issues. It took the scientific community a long time to realize that they were no longer bystanders and that, as members of the society, they needed to be actively involved in the dialogue on these issues.

I am therefore very happy that this book is being published. The book provides a critical overview of the GMO debate and the key issues to be considered. It does so in a unique way by presenting GMO research in a societal setting, showing the intricacies and interdependencies of science, risk management, decision making, stakeholders, and social responsibilities. The emphasis on interdependencies, yet with a clear segregation of discussed topics, provides the needed to unclutter the GMO debate. The lessons learned should be applicable to any new scientific application or technology.

Prof. Gerhard Wenzel
Technical University Munich
Chair of plant breeding

Acknowledgements

We would like to express our gratitude to Visions Unlimited Medien GmbH for providing the CD format of their film "Genes on the menu" adding value to the book. Our thanks also to Dr. Vasanthi Vanniasingham for reviewing some of the more difficult book sections. The preparation of the book content was supported by the European Commission, project number HPRP-CT-1999-00010. The views expressed in the book are purely those of the authors and contributors and may not in any circumstances be regarded as stating an official position of the European Commission.

Dr. Paul M. Pechan   Gert E. de Vries

List of Contributors

Ervin Balazs
Agricultural Biotechnology Centre (ABC), Godollo, Hungary
e-mail: balazs@abc.hu

Anne-Katrin Bock
European Commission,
Joint Research Centre-Institute for Prospective Technological Studies (IPTS),
Seville, Spain
e-mail: Anne-Katrin.Bock@ac.en.int

Andrzej Czaplicki
Plant Breeding and Acclimatzation Institute (PBAI), Blonie, Poland
e-mail: a.czaplicki@ihar.edu.pl

Jaroslava Ovesna
Research Institute of Crop Production (RICP), Prague, Czech Republic
e-mail: ovesna@vurv.cz

Paul Pechan
Ludwig Maximilians University Munich, Germany
e-mail: ppechan@compuserve.com

Giorgos Sakellaris
National Hellenic Research Foundation, Institute of Biological Research
and Biotechnology, Athens, Greece
e-mail: gsak@eie.gr

Gert E. de Vries
Pro Bio Partners, Overschild, The Netherlands
e-mail: pbp@bart.nl

Janusz Zimny
Plant Breeding and Acclimatzation Institute (PBAI), Blonie, Poland
e-mail: j.zimny@ihar.edu.pl

Contents

Introduction

Through the centuries, farmers and (more recently) scientists have searched for ways to improve the food we eat by breeding new, better crops. The driving force behind these improvements was the need to reduce the risk of famine and to increase the stability, wealth and status of a particular nation or region. However, traditional plant breeding has only significantly improved our food supply over the last century. This is primarily due to advances in genetics: the study of how genes are inherited and how they affect the characteristics of living organisms. It is now possible to isolate genes and transfer them between unrelated plants. This technology is called "genetic engineering". It allows the transfer of desirable characteristics from one unrelated organism to another. Products from genetically engineered crop plants have now been on the market for around a decade, with many more likely to come in the future.

In the public arena, genetically engineered plants are most often called "genetically modified", or GM crops. They are often more broadly referred to as "genetically modified organisms", or GMOs.

Many people find it difficult to decide whether GM crops and their products are potentially beneficial or damaging to mankind and the environment. What is different about GM crops and food? Who decides whether they are really safe? What about moral and ethical considerations? What about the impact on our society and trade? What does the public really think about them, and how is public opinion formed?

This book will guide you through the various issues associated with GM crops and food in a systematic manner with emphasis on Europe. It is divided into five thematic sections:

1. Background information on GM crops and their products
2. The science of making GM crops
3. Regulation, assessment and monitoring of GMOs
4. The socio-economic aspects of the GMO debate and decision-making about GMOs
5. The future of GM crops

Before diving into the book, we recommend that you first watch the film "Genes on the menu", which can be found on the CD enclosed with the book. It serves as an introduction to the various issues we will be discussing. The five sections can then be read in any order, depending on the preference of the reader.

The debate about GMOs has been absolutely fascinating, not only due to the important issues it has raised, but also because through this debate we get a chance to observe how our society interacts and evolves. We therefore hope that the book not only provides a thorough source of information on GMOs, but that it also sheds some light on the intricacies and the implications of communication and risk/benefit decisions in our society.

Paul Pechan
Gert de Vries

1 Background on GMOs

1.1
Basic Facts About GM Crops

Paul Pechan

Genetic engineering of crops is a new addition to traditional plant breeding. It is part of our quest to grow more and better food for our growing population. The total area under genetically modified (GM) crop cultivation has been steadily expanding with, however, a marked slowdown in the last few years. Soybean, corn and cotton are the most extensively cultivated GM crops. Europe grows less than 0.5% of the world's GM crops, mainly because of the de facto moratorium imposed on GM crops in Europe until 2003 and the refusal of the European consumers to buy GM products. Currently grown GM crops mainly benefit the farmer, not the consumer. Future uses of GM crops may bring more direct benefits, such as improved taste and quality, to the consumer. All food, whether GM or non-GM, might have certain risks of containing allergic or toxic compounds and all novel foods should be tested for possible allergic and toxic effects on living organisms. Some environmental groups are also concerned about the effect of GM crops on the environment. Most of all they are worried about the unknown effects these crops may have on our environment, and the difficulty of controlling GM crops once they are released into the environment. The discussions about GM crops and food are complex and include issues of choice, globalisation and alternative ways to grow crops. Europe is going ahead with comprehensive labelling and tracing of GM crops and their products. However, even if there were such standards, there is currently no mechanism to implement and enforce these standards on the international market. This first overview chapter is based on the film "Genes on the menu", enclosed as a CD with this book.

1.1.1
Where Does Our GM Food Come From?

The total global area of GM crops exceeded 60 million ha in 2002. In the developing countries, more than 15 million has are under cultivation with GM crops. USA growth nearly 70% of the total, followed by Argentina, Canada and China (for more informations, see Sect. 2.4.1).

Soybean, maize and cotton are among the GM crops that are most extensively cultivated. Soybean represents over 60% of the total GM crop area. These crops were developed to allow better management of weeds and pests. They also include foods like tomatoes and potatoes. In Europe, however, it is not currently possible to buy fresh GM produce. Even in North America, most fruit and vegetables are not yet available as GM products.

Europe grows less than 0.5% of the world's GM crops, mostly maize for animal feed. The European Union has strict limitations on the planting of GM crops. GM crops that have been authorised can be grown, but apparently there is no market to sell EU-grown GM crops. The challenge of coexistence with organic and conventional farming complicates the issues further. European countries do, however, import GM crops and food products from abroad. Thousands of tons of GM soybean seed are imported to Europe as feed for livestock. Soybean also enters our markets as part of many processed food products. Background on current GM crops can be found in Sects. 2.1 to 2.4.

The European Union has put in place new rules to govern the use of genetically modified organisms (GMOs), including GM crops. When foods contain, consist of, or are produced from GMOs, they have to be labelled as such. However, a provision is made for accidental contamination of food sources by GMOs. In such cases, conventional food destined for human consumption is labelled as GM only if it contains more than 0.9% GM ingredients. Thus food products containing less than 0.9% of GM ingredients do not have to be labelled as GM food. As soybean is used in many processed foods, it is likely that much of the food we eat today already contains some GM ingredients although it is not labelled as GM food. In addition, food originating from animals fed GM crops does not need to be labelled as transgenic. The GMO debate revolves around many inter-linked issues and perspectives. All must be taken into account at a decision-making level. Nevertheless, GMOs are a reality of today's market. The need is thus for traceability and labelling to ensure safety and choice, as indeed has been proposed and enacted in Europe. However, labelling does not say anything about the safety of the product: if the product is not safe it will not be allowed onto the market. It also gives consumers the choice of deciding whether or not to purchase GM products for personal use. The main purpose of tracing and labelling GM food in Europe is to identify the precise origin of the food product, so that if a problem arises it can be dealt with quickly. Traceability and labelling of GMOs will cost millions of euros per year. Some propose it may be better to have the regulatory authorities simply decide whether the foods are safe or not and allow retailers to choose whether to label or not. For information about GM product traceability and labelling please see Sects. 3.3 and 4.1.

1.1.2
What is Different About GM Crops and Food?

Man has been modifying plants for over 10,000 years. In plant breeding, a large number of genes are mixed by crossing different plant varieties, and after subsequent screening, plants with the most desirable characteristics are selected. In genetic engineering, only one or a few selected genes from other organisms that have been previously studied are added to a specific living plant cell that can be regenerated into a whole plant (see also Sect. 1.2).

Potatoes are a good example of our breeding efforts. Over the centuries, traditional plant breeders have turned the wild potatoes still found in South America into the wide range of potato varieties we can buy today. Because of these efforts, potatoes can today be grown around the world. These potato varieties are very different from their wild cousins. They have been genetically modified through traditional plant breeding. Every variety of potato differs in its genetic material, the DNA. Sections of DNA form genes that give instructions how to make different proteins. It is primarily the proteins that determine the exact properties of the plant.

A genetic engineering procedure where a gene from one, sometimes unrelated, organism is transferred to another can give a plant a new property, like resistance to insect or virus attack and tolerance to herbicides. The Bt potato, for example, contains a gene from a soil bacteria, *Bacillus thuringiensis*, that makes potatoes resistant to the Colorado potato beetle.

Thus introducing new genes, using the tools of genetic engineering, is a targeted approach for improving plant characteristics. The use of these genes, or transgenes as they are sometimes called, should be a more predictable approach than traditional plant breeding where thousands of unknown genes are exchanged. Unlike traditional plant breeding however, the approach is new, can involve gene transfer that does not occur naturally and has not been tested for long-term side effects.

1.1.3
Why do Some People Worry About GM Crops?

When dealing with new technologies, it is not surprising for the public to be concerned and feel that they do not have much control over the new developments.

Results from the Eurobarometer (55.2, Spring 2001) dealing with public perception of GM food, showed that at least two thirds of those surveyed thought GM food was dangerous to eat, regardless of their education, sex or age. A quarter of the population was unsure about the dangers. Interestingly, the opinion was almost evenly split when asked whether the media over exaggerated the dangers. However, nearly 70% of people asked did not want to eat GM food. Almost everyone wanted the right to choose whether or not to eat GM food and agreed it should be introduced onto the market only when scientifically proven safe.

The implications of these results are that:

- The public overestimates their knowledge of GM food (current GM food is safe to eat).
- Although education did somewhat increase acceptance of GM food, the % of people who rejected GM food was surprisingly similar regardless of their educational background.
- The public seems to be misinformed about the procedures of food approval. All GM products have to pass though extensive scientific testing procedures. Yet the public on one hand feels that GM food is not safe and on the other hand would accept GM food on the market if proven to be scientifically safe.

– It seems that there remains an information and communication gap between
  the public, risk managers and decision makers. This gap has most likely arisen
  because the decision-making process is not fully transparent to the public
  and the public does not trust the process of getting the products onto the
  market.

The public may thus perceive new biotechnologies to be riskier than they actually are. In addition, the European public mistrusts governmental institutions to
protect the interests of the public and feel that it is being left out of the decision-
making process. These conclusions may be, however, specific for Europe and be
to some extent related to the public's experiences with institutional handling of
food-related scandals such as the BSE outbreak and other recent food crises in
various European countries. The issues of public perception of GM crops are
dealt with in Sects. 4.3 to 4.6.

Another public concern is the problem of intellectual property rights and
monopolies. Some people are worried that parts of our common heritage may
become the private property of an individual or a large company.

1.1.4
Why Use GM Crops?

Current usage of GM crops mainly benefits the farmer. Insects and weeds
can do great damage to crops and thus negatively affect the income of a farmer.
For example, the Colorado potato beetle can, within days, destroy a potato
field and endanger the livelihood of a farmer. GM Bt crops have an in-built
defence mechanism against this pest, where the protein made from the inserted
gene disrupts the insect's digestive system. The advantage of planting Bt crops
is that it can reduce the use of insecticides. This means less runoff into the
environment where insecticide pollution can cause damage to other living
organisms.

The claims that Bt crops and useful insects will be killed in greater numbers
than with other treatments have so far not been substantiated, although resistance
to pest control is acknowledged to be a question of time. Refuges containing non-
transformed plants must now be planted around GM crops to reduce selective
pressures for development of pests resistant to, for example, Bt toxin.

Future uses of GM crops may have more direct benefits for our society: new
foods may be created that are better-tasting, contain specific ingredients to
enhance our health, or manufacture life saving compounds. These functional
crops will differ significantly in content and metabolism from the original
non-GM plants and will need to be extensively tested prior to general release.
Details of current and future benefits of GM crops are discussed in Sects. 2.4, 5.1,
5.2 and 5.3.

1.1.5
How Does the Body Deal with GM Food?

Our bodies use exactly the same processes to digest GM and non-GM food. We
eat millions of molecules of protein and DNA at every meal. What happens when

we eat a potato chip made from a pest-resistant potato? It is the job of the digestive system to release molecules from the food we eat. Once inside the body, the potato chip, like any other food, is gradually broken down into smaller and smaller pieces by digestive juices. It is in the small intestine where many of the molecules released from the potato chip are absorbed into the blood. The DNA, which contains the instructions for the synthesis of a specific protein, and which also makes the GM potato pest resistant, is treated the same way as all the other potato DNA.

However, certain proteins in all foods can be a potential health risk. This is true both for GM as well as non-GM food. One way of avoiding health risks like an allergic or toxic reaction is to test foods before they come to the market. Indeed, understanding safety issues, their definitions and limitations is in itself complicated (see below and Sect. 3.5, 3.6 and 4.5).

1.1.6
How are GM Foods Tested?

Initial work growing genetically engineered plants is carried out in a laboratory where the researchers must follow strict safety rules set out by government authorities. GM plants are grown to full size in growth chambers and contained greenhouses under conditions where they can be controlled and monitored. When scientists and authorities are satisfied with the test results, small scale planting can be carried out in the field. These so called enclosed plot trials usually run for 3 years and are followed by large scale open field experiments. The GM product can enter the marketplace only if it complies with appropriate food regulations. It takes approximately 10 years from the beginning of the first experiment until the transgenic cultivar is marketed and sold to the farmers.

The European novel food regulation, for example, requires GM crops and their food products, as well as all other foods new to the European market, to go through extensive safety testing procedures. However, scientists know very little about the risks associated even with non-GM food. It is possible to argue, on the basis of experience and selection, that today one can consider the non-GM food we eat as being generally safe. With GM food, it is not possible to make such a statement as the products have been on the market for only a short period of time.

GM foods are analysed to make sure they do not contain substances that might cause toxic or allergic reactions. Toxicologists play an important role in this part of risk assessment. If a toxicologist is uncertain of the health consequences of a novel food being placed on the market, animal or human trials may need to be carried out before the product is released to the public. Tests like these help experts decide whether foods are safe for the consumer. Insect-resistant and herbicide-tolerant crops have been tested this way before they passed as safe for human consumption. All the currently marketed GM crops and food products can be said to be substantially the same (equivalent) as the non-GM controls.

Experts are now concerned with deciding whether or not to allow onto the marketplace second and third generation GM food products, where plant metabolic processes are manipulated,. In these products, the metabolism of a plant is changed to such an extent so that it is no longer substantially equivalent to the original, non-GM plant. For more information on the topic of regulating food safety, please see Sects. 3.3, 3.4 and 4.2.

1.1.7
What are the Effects of GM Crops on the Environment?

The public has four main concerns: that GM crops will create superbugs, super-weeds or reduce biodiversity, but above all, they are worried about the irrever-sibility of GM crop releases. They are concerned about the environment in general, but are generally not opposed to using genetic engineering in self-contained facilities. They are primarily worried that in the open environment pollen from transgenic crops can pollinate other related plant species and result in unforeseen permanent effects on the environment and biodiversity.

Pollen is carried mainly by wind or insects. The distance pollen can be carried varies with the size of the pollen, wind conditions and how far insects can fly. Using the potato as an example again, in Europe and North America there are no wild relatives of the potato. In South America, where potatoes originate, the situation is entirely different. There, potatoes could cross with their wild relatives. After crossing, these plants would produce seeds containing the new genetic information. New genes could then spread to the wider population. We must take great care not to add any new genes to local plant species because in these centres of origin, the local plants serve as the genetic pool to create new plant varieties. This may require setting up buffer zones, where planting of crops (GM and non-GM) related to the wild relatives, is prohibited. The same is true for example of maize that originates in Central America.

Even outside the centres of origin, great care needs to be exercised when releasing GM crops into the environment because there is a danger of transgenes drifting from one plant population to another. This has happened in Canada with rapeseed and has led to accumulation (stacking) of various herbicide-resistance genes in one rapeseed variety. The concerns about gene "pollution" holds true, however, for both GM and non-GM crops. Thus if cultivated crops, both GM and non-GM, cross with their wild relatives, there could be the danger of creating superweeds resistant to herbicides.

Now, because we can trace individual agronomically important genes in GM crops, the study of gene persistence in wild populations has expanded sub-stantially. The assumption at present is that genes that are valued in cultivated crops will not likely persist in wild populations if they are not seen as an ad-vantage to the wild population. For more information on this topic please see Sect. 2.5.

1.1.8
How are Public Opinions Formed?

Different interest groups use a variety of tactics to get their viewpoints into the public arena, either for or against GM crops. The way the public sees plant biotechnology is often determined by images of activists, dressed in white protective suits, destroying crops. Many people remember this as something dangerous and connect it with the new technologies. Even if people knew the full story, they still may have concerns. The majority of people in Europe are uneasy about GM crops.

For many people, attitudes to new technologies are formed in part by their inner moral beliefs. Here, the views of the church can be important. Interestingly, the Catholic church is not against GMOs per se. The Pope and the Church have not stated that GMOs are intrinsically bad or that they should not be grown. They believe a great deal of caution should be exercised to be sure that there are no risks for human health. There has not been any theological condemnation of manipulating living organisms (apart from genetically manipulating mankind itself). The views of other religions or groups may differ depending on the circumstances. For example, if a gene from a pig is introduced into a plant, certain religions may reject the food as unclean. However, the issues of GMOs have transcended moral and safety considerations. They have become part of the social and globalisation, including monopoly, issues, and debates. Sections 4.3, 4.4 and 4.6 address the topics of public opinion, opinion formation and moral attitudes.

1.1.9
What GM Products Can be Expected?

The first commercial products of genetic engineering were plants resistant to pests and weed killers. It is mainly farmers that reap the benefits. Now, researchers in Europe and many other parts of the world are developing new products designed to be of direct benefit to the consumer. These are the so-called second and third generation GM products. The development of new products is driven by industry, available methodologies, scientific curiosity and social need. Areas under active research include plants that are resistant to high salt content in the soil or to drought. To succeed, researchers need to add several new genes to the plant. Other research involves trying to change what happens inside the plants themselves: their metabolic pathways. Major efforts are going into creating foods designed to improve our health or to prevent diseases. For example, tests are underway on a GM rice designed to make a vitamin A precursor to supplement the diet of people deficient in vitamin A. Many of these research advances may be used in the developing world. Future uses of GM crops will be for what is termed molecular farming – using plants as bioreactors to produce, amongst others, medical products and GM foods that directly benefit the consumer in terms of price, taste, health and extra quality. For more information, please see Sects. 2.1 to 2.4 and 5.1 to 5.3.

1.1.10
Can GM Crops Help the Developing World?

For developing countries, research decisions depend on local priorities and are often a question of need rather than choice. Many companies do not consider poor developing countries as a target for their products, nor would they develop products just for these markets. Indeed, it is not the mandate of companies to help the developing world. In these smaller and poorer markets most product development is carried out by public institutions. There, the trend is towards the production of GM crops with improved quality traits such as salt or drought tolerance to improve the crop yields or increase the use of low quality land for cultivation. In other parts of the world, pest control is a major problem. Using chemicals to control pests may create risks to wildlife and to human health and is relatively expensive. Poor instructions along with insufficient training are the main problem. These problems, combined with poor storage conditions, can cause major pollution. Chemical contamination of land is a major tragedy in the developing world. Pest-resistant plants are one way of cutting down on the use of chemical sprays. Genetic engineering, however, cannot offer a complete solution to complex problems, whether natural or man-made.

Sometimes, low technology solutions in developing countries can be just as effective or more appropriate to tackle a problem. The solutions can differ from country to country. The solutions can range from better education of farmers, efficient use of water (for example, drip irrigation), better infrastructures to bring produce to the market place, better storage facilities (in some countries more than 20% of the produce is lost during storage) and using local crops rather than planting imported crop varieties. The issue of GM crops and the developing world is addressed in Sect. 5.3.

1.1.11
Possible Alternatives to GM Crops

The overall aim of agriculture today is sustainable land use and sufficient food production. This primarily means reducing energy inputs into producing food while at the same time treating the environment with respect. GM crops are only one approach to achieving sustainable agriculture. Other approaches include conventional agriculture, integrated farming, low-input farming and organic farming. The latter methods are becoming very important for farming as well as being appealing to consumers. In Germany, for example, the aim is to have 20% of agricultural land under organic cultivation within the next decade. Organic farming is based on respect for nature and avoiding stress situations for plants by balancing the needs of microorganisms living in the soil with those of the plants and the insects that feed on the plants.

Instead of chemicals, organic farmers use biological methods to control pests. A common way is to release natural enemies of the harmful insects. Bt toxin and other pesticides that are extracted from natural materials can be used by organic farmers. In GM crops, the active ingredient of Bt is produced by the plant itself. However, organic farmers object to the use of GM crops as being unnatural.

The decision as to which agricultural approach to use (whether GM, conventional or organic), should depend on comparing the alternatives for a specific area and the objective. This analysis should be based on a detailed risk–benefit analysis that includes health, environmental, social and economic dimensions.

The best alternative, or combination of alternatives, should be used to solve a given problem.

The decisions concerning what and how to farm are thus influenced by many factors, some preventing the combination of various approaches, others encouraging them. It is up to the decision makers and governmental organisations to encourage, monitor and enforce the proper agricultural practices. The topic of coexistence of GM and organic farming can be found in Sects. 2.5, 3.5 and 4.1.

1.1.12
Conclusion

One problem for decision makers is that there are no agreed upon international standards for assessing the risks of GM crops and GM food. Different countries base their judgements on different sets of questions and not all countries have access to good testing facilities.

In the USA, decisions are based primarily on scientific judgements about safety. This approach means that GM developments are moving ahead quite fast but if something goes wrong, guilty parties may be sued. This, combined with ensuing negative publicity, functions partly as a mechanism to ensure "honesty" in their business community. Europe's precautionary principle tries to pre-empt unwanted consequences. The precautionary principle is a decision-making tool that helps to arrive at decisions in the absence of hard scientific data. This allows decision makers to proceed cautiously with GM crops and food. Their decisions are based on social considerations as well as the available scientific evidence. Without an international agreement on standards, food producers face real problems in developing and selling GM products to countries in other parts of the world. Even if international standards can be agreed upon, there is no mechanism to enforce them. This is the real challenge for the decision makers. These complex issues are discussed predominantly in Sect. 4.2.

1.1.13
Information Sources

Parsley GJ (2003) New genetics, good and agriculture: scientific discoveries-societal dilemmas. International Council for Science. The full text can be downloaded from http://www.doyle foundation.org

Visions Unlimited Medien GmbH, Genes On the Menu. For more information, email: VUMedien@AOL.com. A CD version of the film is enclosed with this book

1.2
History and Uses of Plant Biotechnology

Andrzej Czaplicki, Jaroslava Ovesna, Gert E. de Vries

The improvement of plants for food production and the use of conservation methods have been in practice as long as humans settled down to rely on agriculture for sustenance. The techniques and successes of traditional plant breeding have, in the past, evolved gradually. With the advent of genetics, the advancements have accelerated and allowed us in the 1960s to greatly increase the productivity of our land. This enabled many countries, for the first time, to provide adequate food supply to their growing population. Essential knowledge and understanding of cell function and heredity combined with new possibilities to modify and transfer DNA between organisms is only a few decades old. These advancements have resulted in the development of efficient vaccines and pharmaceuticals, new food technologies and many other products improving the overall standard of life. This is also true of agriculture where genetic engineering of crops can complement traditional plant breeding to suit the needs of today's society. Most of these advances can be grouped under the term "biotechnology", which aims to use organisms, cells or part of cells in technical/industrial processes.

1.2.1
Traditional Biotechnology

When humans realised that they could grow their own crops and breed animals with improved characteristics, they effectively started to use biotechnology. The first biotechnological processes that were recorded in a written form concern food production around 5000 BC. Our more recent ancestors were producing bread, wine and beer and used fermentation, all natural microbiological processes, in part to prevent foodstuffs from spoiling, in part for health and pleasure. Many of our current food products are still using the same methods of fermentation that were discovered thousands of years ago. Plant species have been improved by selection and crossing to obtain more yielding varieties with better nutritional traits.

1.2.2
Modern Biotechnology

The roots of modern biotechnology can be dated back to the work of Louis Pasteur, Robert Koch and Gregor Mendel, approximately 100 years ago. Pasteur and Koch founded the basis for the current science of microbiology. When Pasteur discovered the efficacy of weakened rabies virus to prevent the incidence of the same disease, the technique of preventing infection diseases by means of vaccination was established – having been pioneered by Edward Jenner in the case of smallpox. Mendel described the laws of inheritance, the transfer of parental properties (genes) to offspring. It was Karl Ereky in 1919 who used the term biotechnology for the first time in order to describe the interaction of biology and human technology. Thus, one definition of biotechnology is that it is a method through which life forms (organisms) can be manipulated to provide desirable products.

A common misconception today is that biotechnology refers only to genetic modification (also referred to as genetic engineering, genetic manipulation, gene splicing, recombinant DNA or gene technology, etc.). Genetic modification, however, is only one of many techniques used to derive products from micro-organisms, plants or possibly animals that may be used in the biotechnological industry. A list of areas covered by the term biotechnology would more properly also include plant tissue culture, mammalian cell culture, enzyme systems, plant breeding, immunology, fermentation and others. In this chapter the term "modern biotechnology" will be used when speaking about techniques using genetic modification.

Genetic modification can be defined as the technique that involves the isolation of genetic material, splicing, altering, recombining and transferring it from one organism to another. These techniques can be performed at various levels: from whole genome manipulations through chromosome manipulations to precise modification of single genes. Genetic modification has come to include the manipulation and alteration of the genetic material of an organism in such a way as to allow it to produce proteins with properties different from those normally produced, or to produce entirely foreign proteins altogether.

It is now a routine practice to combine, exchange or mix genetic material. If thought useful, DNA can be isolated from one organism, combined with DNA from another (so called recombinant DNA) and placed into cells of a third organism. As a result, it is for example possible to have bacterial cells produce human proteins, such as human insulin. Therapeutic compounds (e.g. interferon, inter-leukin) that were unavailable previously can now be synthesised in a variety of systems, ranging from micro-organisms to plant or animal cells. Biotechnology can be also used to produce artificial transplant materials such as blood-vessel or skin tissue. Basic knowledge, as well as novel pharmaceuticals from biological origin, contribute significantly in treatments for cardiovascular diseases, cancer, stroke, etc. Biotechnology is also applied in other areas such as the environment and food. Micro-organisms have been used to efficiently extract minerals from ores and to clean up pollutants. Genetically improved micro-organisms are being used in the food industry, and give us yoghurts, beer or cheese with improved texture, taste or quality. It is clear, that biotechnology contributes to many fields of man's activities. In agriculture, it has brought us a range of methodologies for high quality plant breeding that result in both fast results and superior varieties.

1.2.3
Plant Breeding and Plant Biotechnology

Almost without exception, crop plants in use today have been adapted to the needs of mankind. The development of new plant varieties through selective breeding has been improving agriculture and food production for thousands of years. The genetic make-up of crop plants has been changed by mankind's selection of naturally occurring variants. What has come to be called biotechnology and the genetic enhancement of agricultural products may be one of the oldest human activities. Traditional plant breeding involves crossing of different plants with useful characteristics, and this has been very useful in improving crop plants. Since the entire gene pool of each of the two partners is mixed and useful trait(s) must be selected upon successive rounds of crossing and selection, plant breeding is a time-consuming process. Improving crops through traditional methods is also subject to restrictions imposed by sexual compatibility, which limits the diversity of useful genetic material.

Undoubtedly however, classical breeding has dramatically increased the productivity of the plants we grow for food, fibre and other purposes. The so-called Green Revolution during the years 1950–1965 involved the simultaneous effects of breeding efforts and altered agricultural practices. New crop varieties and the availability of pesticides and fertilisers greatly increased the needed crop yields in developing countries. Mexico, for instance, was importing half of its wheat consumption in 1944. In 1956 the country was self-supporting, and by 1964 Mexico was exporting 0.5 million tons of wheat. Apart from crop yield, significant progress was made in improving the quality of plant products such as proteins, carbohydrates, fibre or oil content.

Technological developments, which allowed the manipulation of plant cells, were a major requirement before plant biotechnology really could take off. Although it had been shown around 1900 that it would be possible to regenerate an entire plant from just a small part – omnipotency – it would take many years before this finding could be put to practice. Efficient in vitro cloning and propagation techniques, sterile ways to reproduce identical, healthy plants of valuable genotypes in tubes, were only recently established for most plant species. The technology can now be used to maintain disease-free plants prior to large-scale production. Ornamental plants, high quality banana seedlings or potato microtubers are produced using these techniques.

Although still a relatively young scientific discipline, genetic modification is being applied in a broad variety of ways, mainly in biomedical research, but also in agriculture and food production. With the help of genetic modification, it is now possible to transfer a gene from a bacterium to a plant so that the transgenic plant produces the corresponding bacterial protein in its cells. Thus genetic modification enables the targeted transfer of individual traits that are encoded in genes to a given plant, allowing the transfer of only one or a few desirable genes to develop crops with specific traits. Genetic modification is partly seen as complementing and extending traditional plant breeding. Others perceive it as a completely new way of generating plant varieties.

1.2.4
Plant Biotechnological Techniques

While the DNA molecule (deoxyribonucleic acid) from a micro-organism, a plant, an animal or a human is chemically identical, the sequence of its building blocks differs significantly. Genes are units of sequences, which bear resemblance in different organisms, and their cohesion and collaboration is the basis of life. A gene is a blueprint for the synthesis of a specific protein. The genetic code in the blueprint is universal, therefore any gene can be made understood in each organism. This is the basis for genetic transformation, whereby genes can be taken from any source (plant, animal or microbe) and inserted into another organism where it can be expressed as a normal part of the genome. Plants that were made resistant to herbicides or produce advanced pharmaceuticals serve as an example.

Advances in genetic modification are based on a long history of scientific enquiry. One of the key events was the discovery of the fundamental principles of heredity by Gregor Mendel at the end of the nineteenth century. He established the basic laws of genetics that led to our understanding of the inheritance of traits and the role of genes in transferring these traits to offspring of plants and animals. With the understanding of the laws of heredity, plant breeding protocols were devised whereby selection was accompanied by deliberate crossing. Subsequent investigations advanced our understanding of the location, composition and function of genes.

1.2.4.1
Genetic Markers

Mendel described the laws of heredity following his pioneering work with pea plants. While his rules for the segregation of traits are generally valid, exceptions can always be observed in the form of enhanced linkage of genes. Later, such groups of genes were found to be located on the same chromosomes and as a consequence have a greater chance to be inherited together. These linkages provided breeders with the possibility of detecting the presence of one gene by the detection of a neighbouring gene, a genetic marker. Easily detectable markers, such as flower colour or seed shape, have facilitated the tracking of other, more complicated, traits such as stress resistance or crop yield in breeding programmes. Four types of genetic markers can been used, which can be divided according to the method of analysis that is based on morphology, cytology, protein chemistry or the detection of DNA sequences. Such genetic markers therefore function as indicators for the possible presence of linked, and desirable, stretches of chromosomal DNA or individual genes.

A morphological marker is a trait that can be seen directly in plants growing in the field, examples are flower colour, aberrant stem length, leaf shape etc. The choice for such a marker has two serious drawbacks: the number of useful morphological traits is limited, and the phenotype itself is often undesirable. Cytological markers rely on regions of the chromosome, which can be preferentially stained to reveal the presence or absence of certain sequences in the genome. Unfortunately, cytological techniques are time-consuming and too inefficient to be used routinely in a plant-breeding program. The use of protein markers relies on the ability to detect specific proteins in extracts from plant tissues. This method is relatively inexpensive and efficient but the number of known and useful markers is very limited.

DNA markers offer the greatest potential. In 1980, it was suggested that such markers could be identified by fragmenting the DNA with restriction endonuclease enzymes (enzymes that can cut DNA at specific locations) and pin-pointing individual traits to specific fragments. Restriction fragment length poly-morphism (RFLP) became an important technique in the analysis of genomes from microbes to humans and has been used extensively to develop genetic maps. The practice has now been revised by the use of a new technique, the poly-merase chain reaction (PCR). This method is based on the ability to detect DNA segments through amplification by a process of DNA synthesis. Various genetic marker systems for crop improvement have now been developed using PCR-based strategies. The technique is almost ideal because genetic markers can be developed for most regions of the genome and they are highly polymorphic (variable), which means that even individual gene variants can be tracked. Since the technology involved is simple, the use of molecular genetic markers have become routine in most breeding stations and an established tool in plant breed-ing. The molecular techniques, although they are used in the process of genetic modification as well, are themselves no subject of public opposition since they are only a tool to accelerate plant breeding and selection.

1.2.4.2
Manipulation of Plant Genomes

Plant genes are organised on chromosomes, which are normally located in the nucleus of plant cells as sets of pairs to form the plant genome. The paired chro-mosomes originate from the parent plant(s), one from each parental diploid set. The number of chromosomes in a plant cell varies from species to species, and the whole genome may have undergone multiplication, leading to polyploidy. Polyploidy plays a major role in evolution and has also been shown to improve certain characteristics of cultured plants. Geneticists have therefore increased the ploidy level of certain crops to achieve greater vegetative production, larger flowers and improved seed. More than one third of all cultivated plants in use are now polyploid.

Some plant species are natural hybrids between two separate species and thus carry two genomes, each of which originates from a distinct diploid species. Plant breeders have used such anomalous genetic processes and, in their search for the combinations of best qualities, produced new plant genome combinations through interspecific crosses. In the 20th century, scientists succeeded in the production of a wide range of hybrids by the use of tissue culture techniques. Using these techniques it also became possible to generate large numbers of genetically identical plants (clones) in a period of only a few weeks or months. Somatic hybridisation is a more advanced method that is used in the laboratory to produce interspecific hybrids between plants that normally would not cross or produce viable offspring. The technique relies on the fusion of plant cells lacking cell walls, called protoplasts, followed by growth in tissue culture and plant formation. Somatic hybridisation offers the possibility to introduce genetic material from non-compatible wild species into cultivated species.

1.2.4.3
Chromosome Manipulations

Evidence has been found that some species evolved through interspecific crosses and hybridisation between species that possess dissimilar genomes. Viable (sometimes barely) offspring may be produced with more or fewer and rearranged chromosomes than the present in the parental species. Aberrant plant genomes may also arise when complete chromosome(s) of one species are replaced with equivalent chromosome(s) of another variety or related species. The instability of the resulting plant genomes increases with degree of evolutionary distance of the parent plants.

The deleterious effect of the loss or the gain of a full chromosome carrying unwanted as well as necessary genes, can be overcome by natural rearrangements, breakage or deletion of segments in such a chromosome. Chromosome breaks may occur naturally at meiosis, but in the hands of a researcher they can also be induced by certain chemical and irradiation treatments. Broken chromosomes have the ability to fuse with each other to produce translocated, or rear-ranged chromosomes, which, with luck, may provide the plant with improved characteristics. Although advanced techniques are now available to introduce whole chromosomes or chromosomal segments in plant cells, the limitation of this strategy (as is also true in conventional plant breeding) is the laborious procedure to select and further develop desired genotypes from the progeny.

1.2.4.4
How Can Genes be Altered?

So far we have discussed the evolution of crop improvement from domestication to deliberate plant breeding and how overall, uncontrolled, genetic changes can be achieved. Genetic improvements have long been based on simple selection in the field. Relatively recent efforts combine planned crosses between defined parent plants, combined with a targeted selection for defined improvements. These processes rely on natural or induced recombinations of genetic material to reshuffle genes. A major limitation in this approach is that the desired trait or phenotype must be present in the parent plants that can be crossed. What if a desired variation does not exists among the available plant varieties?

1.2.4.4.1
Mutation

In 1901 Hugo de Vries published his mutation theory, arguing that progressive mutation could bring about change within a species. While many mutants had been produced in a number of crop plant species, most had changes for the worse. Moreover it was found difficult to select those mutant plants which had acquired improved characteristics. Several thousands groups of plants had to be screened to find a rare desirable type and this was considered too random a process for the time. Today there are new methods of inducing mutations and the improvements in selection and detection (such as the use of genetic markers) has given new impulse to mutation-based breeding strategies. The power of the selection of mutations for breeding purposes can be illustrated by an example of maize lines that were mutated to gain resistance to a herbicide compound. By itself this an unremarkable event since such lines had been produced before in other crops. But a remarkable finding was that herbicide resistance was due to a single amino acid replacement in the enzyme targeted by the herbicide. This example demonstrates that unexpected properties may arise from seemingly trivial changes in the genetic makeup of an organism that cannot be predicted. It is clear that, while a proper method for inducing mutations is important, an essential step is the design of a clever system to select the desired plant lines.

1.2.4.4.2
Transformation

Genetic transformation is a technique whereby genes can be taken from a selected organism (microbe, plant or animal) to be introduced into another organism where they can be expressed as a normal part of the genome. The first step in this process is to splice a selected gene segment from the DNA sequences of a donor, using restriction enzymes. This DNA fragment is then linked to other segments of DNA that contain marker and selection genes. Several methods are used to actually introduce the prepared donor DNA into cells of the target organism. A no-nonsense method just shoots gold particles, coated with the target DNA, at random into a mass of cells. In a only few cells will what is aimed for happen: the donor DNA is taken up and incorporated into DNA material of the target organism. Thanks to the presence of selection genes, successful transformed cells will grow and stay alive on a growth medium that prevents the development of untransformed cells. Subsequent analyses will reveal whether the selected gene functions correctly in the target organism.

The story of genetic modification in plants started 1980, when it was demonstrated that a soil bacterium, *Agrobacterium*, caused tumours in plants after transferring a small but distinct DNA fragment to a plant cell, where it would be incorporated in the nucleus and change the physiology of the local tissue. In 1983 the system was put to use and GM *Agrobacteria* were used to transfer an antibiotic resistance gene into a tobacco plant. Today advanced methods for the genetic transformation in a wide range of plants have been developed that are based on this natural phenomenon.

It should be noted that the term "transformation" is a bit of a misnomer. Most transformation work in crop plants involves the transfer of single genes. The resultant plants are not radically changed, new species are not produced but rather the crop plant is modified by an acquired new trait.

1.2.5
Conclusions

Plant biotechnology can complement traditional plant breeding to suit the needs of today's society. However, it has not often been recognised that plant biotechnology consists of many techniques and that only some require transfer of genes between unrelated organisms. Since the initial successes of genetically modifying plants, a wide range of applications have been investigated to solve existing problems in agriculture such as problems with insects, plant pathogens and weeds as well as abiotic stresses. Edible vaccines for infectious diseases like cholera, hepatitis B, and diarrhoea, when produced by GM plants and administered as a food component, may in certain instances help to efficiently fight diseases in the Third World. GM plants may have a role in cleaning polluted soil (phytoremediation) and may also be suitable as fuels in power plants (see Sects. 5.1 to 5.3).

In spite of the promising contribution of genetic modification of plants, many points of criticism have been raised by non-governmental organisations, activist groups, politicians and individuals. Opponents argue that genetic modification is not required to feed the world or to solve problems of agriculture production. Problems in the world are primarily caused by the economic balance of power, war, drought, mismanagement or poverty and not by lack of resources. Perceived environmental or health risks when using GM organisms have also motivated European citizens to protest against the use of GM plants and biotechnology.

1.2.6
Information Sources

Agrifood Awareness Australia, How plant breeding works. http://www.afaa.com.au/pdf/4-Plant_breeding.pdf

Australian Office of the Gene Technology Regulator, World agricultural biotechnology on GMOs: What Is biotechnology? What is gene technology? http://www.ogtr.gov.au/pdf/public/factwhatis.pdf

Barth R et al, Genetic engineering and organic farming. German Federal Environmental Agency. http://www.oeko.de/bereiche/gentech/documents/gruene_gentech_en.pdf

Conko G, The benefits of biotech. http://www.cato.org/pubs/regulation/regv26n1/v26n1–4.pdf European biotechnology: development and outlook. http://biozine.kribb.re.kr/special/1–3-03.html

European Plant Biotechnology Network. http://www.epbn.org/

FAO Comm Genetic Resources Food Agric, The status of the draft code of conduct on biotechnology. http://www//ext-ftp.fao.org/ag/cgrfa/cgrfa9/r9w18e.pdf Information on biopharmaceutical production. http://www.ul.ie/~biopharm/

James C (2002) Global status of commercialised transgenic crops: 2002. ISAAA Briefs 27 http://www.botanischergarten.ch/UNIDO/ISAAA_Briefs_No._27.pdf

Levy A et al, The dynamic plant genome. http://www/avraham_levy.pdf

Rodemeyer M, Future uses of agricultural biotechnology. Pew Initiative. http://pewagbiotech.org/research/harvest/harvest.pdf

USDA, US-information on GMOs in agriculture. http://www.usda.gov/usda.htm

2 The Science of GMOs

2.1
Herbicide Tolerance

Janusz Zimny, Gert E. de Vries

From the very beginning of agriculture farmers have had to fight weeds in order to prevent them from overgrowing cultivated crops. In the nineteenth century, chemical substances (herbicides) were discovered that would inhibit or arrest the growth of weeds. Herbicide use became current practice in the last few decades but, while improving crop yields worldwide, some chemicals caused serious environmental problems. Continuous use of herbicides also caused that some weeds adopted herbicide-tolerant forms rendering those herbicides useless. In the 1980s, plant researchers identified and isolated single genes that were responsible for herbicide tolerance. After introduction of the genes into crop species, these plants gained herbicide tolerance. Farmers now could spray their fields with herbicides to selectively control weeds and leave their crops unaffected. Controversies have arisen concerning the use of herbicide-tolerant crops, mainly because of possible escape of the introduced genes through pollen or seed dispersal. However, many benefits have been documented: an increase in productivity, conservation of soil structure and an improved use of herbicides.

2.1.1
Introduction

From the moment people started to employ agricultural methods and grew plants for their own benefit, they have faced the problem of eradication of weeds in order to ensure the best possible conditions for cultivated crops. In the beginning weeding depended on manual labour. Only much later were mechanical methods introduced, to be replaced by chemical strategies in the 20th century.

The chemical approach to fighting undesired plant growth was developed as the result of an overall expansion of the chemical industry in the nineteenth century. The name "herbicide" was given to chemicals that could be used to kill weeds. Some herbicides selectively arrest growth in broad-leaved plants while leaving monocotyledons (plants like grasses, including rice and cereal crops) unharmed. Other broad spectrum herbicides affect growth in all plants (Table 2.1). Herbicides usually kill weeds by interfering with the function of an enzyme

Table 2.1. Overview of herbicides in use

Herbicide families	Examples
Amino acid synthesis inhibitors	Roundup, Rodeo, Harmony Extra
Ammonia assimilation inhibitors	Basta, Liberty
Cell membrane disrupters	Gramoxon, Diquat, Cobra
Growth regulators	Banviel, 2,4-D, MCPA, Stinger
Inhibitors of plant pigment synthesis	Zorial, Balance, Callisto, Command
Lipid synthesis inhibitors	Assure II, Acclaim Extra, Fusillade
Photosynthesis inhibitors	Velpar, Basagran, Buctril, Aatrex
Seedling growth inhibitors	Treflan, Prowl, Sonalan, Barricade

of key importance in the plant cell. Such enzymes preferably catalyse specific reactions in the metabolism of the plant, having no identical counterpart in the animal kingdom.

Herbicide use has dramatically increased crop yield world-wide but it has also caused serious environmental problems by contaminating soil and water, as well as being a health hazard to both humans and animals. The risk of contamination depends on many factors such as the type of herbicide, soil structure and fertility, the level of ground water, crop rotation, cultivation, irrigation scheduling, frequency of herbicide use, weather conditions and others.

Although herbicide tolerance (HT) may naturally occur in some plants, it has proven more efficient to provide crop plants with selected genes to construct genetically modified plants. Indeed such HT crops were so successful that herbicide tolerance was one of the first genetically modified traits to enter the market. The trait allows farmers to spray herbicides, which control weeds, selectively at any time during the crop season and leave the crop plants unaffected.

While there are many new traits and applications of GM in agriculture, a major focus has been on the development of herbicide-tolerant crops. Multinational corporations had found it commercially attractive to develop GM crops with characteristics such as HT to reduce dependence on such inputs as pesticides and fertilisers.

2.1.2
Plant Sensitivity to Herbicides

In general, plants exhibit a range of sensitivities to the herbicides used in agriculture, with some species being tolerant to a particular herbicide. There are several mechanisms by which plants can tolerate exposure to herbicide:

- The plant produces an enzyme, which detoxifies the herbicide
- The plant produces an altered target enzyme, which is not affected by the herbicide
- The plant produces physical or physiological barriers to the uptake of the herbicide by the plant tissues and cells

2.1.3
Transgenic Herbicide-Tolerant Plants

The application of the technique of genetic modification has produced crop plant species that are tolerant to a range of herbicides. The majority of plants, however, have been modified to tolerate either glyphosate or glufosinate. Transgenic HT was first achieved in 1985, when tobacco cells were made glyphosate-tolerant. Glyphosate is an active ingredient of many herbicide formulations, i.e. Roundup. The target enzyme of this herbicide is the enzyme 5-enolpyruvyl 3-phospho-shikimic acid synthase (EPSPS) which is involved in the biosynthesis of amino acids in the chloroplast. Tolerance was obtained by expressing a gene for a bacterial enzyme that rendered plant cells tolerant to the herbicide. Another transgenic herbicide tolerance system confers tolerance to glufosinate (phosphinotricin, Liberty or Basta). The compound is phytotoxic since it inhibits a unique plant enzyme, glutamine synthetase, which is involved in the assimilation of ammonia. The herbicide causes the plant to perish due to the build-up of high ammonia concentrations. A bacterial gene for phosphinotricin-acetyl-transferase, which inactivates the inhibitory function of the herbicide, confers tolerance to glufosinate in GM plants.

Glyphosate-tolerant GMP varieties are now available for a range of different crop species, including oilseed rape (canola), soybean, cotton, maize, sugar beet, wheat, chicory, cauliflower and broccoli. In addition to being useful to farmers, herbicide tolerance has also proved a useful trait to breeders when selecting plants to be used in the production of hybrid seeds. Unlike many other herbicides, glyphosate (Roundup) shows low toxicity to humans and animals, and degrades quickly in the soil. The use of herbicide-tolerant crops in the USA has in fact reduced the use of herbicides and allowed farmers to adopt no-till farming methods that minimise soil erosion and loss of water through evaporation.

Also, a range of crop species have been made tolerant for glufosinate and many of these plants have been grown in small-scale field tests to evaluate performance under field conditions. These include barley, peanut, sugar beet, wild cabbage, chicory, carrot, tall fescue, cotton, barley, tomato, alfalfa, gladiolus, melon, poplar, potato, rapeseed, rice, soybean, sorghum, sugarcane, tobacco, wheat, triticale and maize.

Tolerance to other classes of herbicides has also been introduced into plants by the use of genetic modification and specific examples are shown in Table 2.2.

Table 2.2. Herbicide tolerances in GM crops

Active ingredient of herbicide	Transgenic plant
Chlorsulfuron	Sugar beet, sunflower
Isoxazole	Maize, oilseed rape, soybean
Oxynil	Cotton, oilseed rape
Sulfonylurea	Sugar beet
Sulfonamide	Oilseed rape

New traits with improved qualities are being developed every day, and an example of this progress is a gene that will provide tolerance to an entire class of broad-spectrum herbicides known as protoporphyrinogen oxidase inhibitors or PPOs. The technology will work on a broad range of crops, including maize, wheat, soybeans, rice, canola, cotton, sorghum and sugar beet. PPO herbicides kill plants, either crops or weeds, by blocking a key metabolic process. The new gene replaces the original plant gene to render GM plants tolerant to the PPO herbicide and leave plant growth unaffected.

2.1.4
Herbicide Tolerance for Hybrid Seed Production

The production of hybrid seeds has proven to be a very useful method to combine quality traits from two plant lines in a progeny crop. Male sterility is used to prevent self-pollination and force the production of hybrid seeds in the mother plants. The HT trait comes in handy here when used as a selection marker. The former Belgian biotechnology company Plant Genetic Systems, for instance, developed a system where herbicide tolerance was introduced together with the gene for male sterility. The tight coupling of these two traits facilitated selection and tracking of male sterile plant lines. In the process of hybrid seed production contaminating pollen donors can now be killed after fertilization has taken place, leaving only mother plants to produce the hybrid seed. As a "bonus", growers benefit from the herbicide-tolerance trait in the hybrid seed and may use the herbicide to control weeds without damaging the hybrid crop. In general, plant lines that are marked by the presence of the HT trait can be maintained more easily by breeders by the use of the herbicide and by eliminating any contaminating parent lines.

2.1.5
Distribution of Herbicide-Tolerant Plants in the World

In 2000, 32.7 million ha were cultivated with GM crops. Among the HT plants, soybean was cultivated on 59% of all fields with GM crops while herbicide-tolerant canola covered 6%, herbicide-tolerant cotton 5%, herbicide-tolerant maize 5%, Bt/herbicide-tolerant cotton 4% and Bt/herbicide-tolerant maize 3%.

Between 1998 and 2000, the global share of transgenic plants production changed considerably. The area of land under transgenic herbicide-tolerant soybean increased to 25.8 million ha and showed the dominance of this crop among all cultivated transgenic plants. This was 36% of the total soybean crop grown globally. In 2000, transgenic cotton was grown on 5.4 million ha (16% of global production), and canola covered 2.75 million ha (11% of global production).

Argentina showed an increase from 4.3 million ha of transgenic plants in 1998 to 6.4 million ha in 1999 and 10 million ha in 2000, mainly HT soybean. The same crop was grown for the first time in Uruguay, but on a small area of 3,000 ha only. In the USA, the cultivated land areas for GM maize (Bt, Bt/herbicide and herbicide-tolerant varieties), increased from 8.1 million ha to 10.3 million ha in 1999 and decreased by about 800,000 ha in 2000. The reason for the decrease of GM Bt crop plantings was coupled to the decreased infestation of European corn borer in year 1999 and an expected further decrease in next years. In Canada, herbicide-tolerant canola was grown on 3.4 million ha in year 1999 and a decrease of 600,000 ha was observed in 2000. In Eastern Europe, market production of transgenic crops started in 1998. Herbicide-tolerant soybeans was first grown in Romania in the year 2000 and herbicide-tolerant maize was grown in Bulgaria. In other CEE countries only field experiments have been approved.

2.1.6
Commercialisation of Herbicide-Resistant Plants

The marketing strategy for HT crop varieties that was initially developed by plant breeding companies focused on farmers. While this has been successful for the a number of countries in North and South America, the GM products could not be exported to many other regions of the world. The only herbicide-resistant crops for which the whole regulatory process had been completed in the European Union were varieties of maize and rapeseed. Imports of beans derived from a glyphosate-resistant soybean also have gained EU approval.

2.1.7
Profitability of Herbicide-Resistant Plants

Since the early 1960s the usage of herbicides in agriculture has increased dramatically. While the advantages of using herbicide over mechanical weed control are clear to most farmers, there are significant drawbacks if this increase would continue. There is always a risk of damage to nearby crops, a risk of the development of herbicide resistance in weeds, the risk of contaminants in food products and the risk of damaging the environment when chemicals persist.

The target herbicides for the HT traits in GM plants are modern chemicals with low toxicity for animals and humans. The compounds are quickly degraded in natural environments. It is therefore positive to note that since the introduction of HT soybean in the USA a considerably decrease in the usage of less favourable types of herbicides has been observed. Since herbicide application can be performed at different phases of plant development, when using GM-HT crops, herbicides can be used more effectively when fighting weeds at a later stage of growth. Other advantages of using herbicide-tolerant plants are a reduced requirement for field work, a reduction in the use of machines and soil erosion, and therefore overall reduced costs for crop production. While HT plants may have advantages, are they also economically profitable when seed prices of GM crops are higher?

The initial success of GM crop products on the US market encouraged high expectations among farmers. They hoped for a rise in crop yields as well as a decrease in production costs. The use of HT crop plants has led to a change in tillage technology and thus to the lowered costs of production although yields were not enhanced. Factors other than profitability and convenience also influence market possibilities as well, notably consumer resistance.

Table 2.3. Profitability of GM soybean vs. the conventional one

Crop	Yield t/ha	Seed cost €/ha	Total cost (excluding land/labour) €/ha	Return on land/labour €/ha
GM-soybean	3,295	57	254	320
Soybean	3,430	42	274	322

Source: Working document rev2 (2000) Directorate-General for Agriculture, Commission of the European Communities.

In the USA two varieties of HT soybean are being cultivated. Glyphosate-tolerant soybean crops account for 80% of the GM-HT soybean production. Results of many analyses conducted for these crops show a 3–13% lower yield than unmodified crops. The reason for this difference may be the fact that the plant varieties that were used to construct the GM-HT lines are less productive than current non-GM varieties.

The data in Table 2.3 show that HT GM soybean plants do not seem to improve profitability of production but, as argued before, the use of HT crops is more convenient for farmers and there are positive environmental aspects. It may be concluded that several issues still need to be resolved and that it will be necessary to wait and observe changes in the market place when new GM varieties are introduced. Marketing strategies for GM-HT crop varieties have focused on farmers and the concept of a "technological package" consisting of the GM seeds plus herbicide formulation. Now it is clear that both the consumers and politics have been ignored.

2.1.8
Controversies about HT Crops

Non-governmental organisations concerned with natural habitats, biodiversity and ecology, organic agriculture or consumer interests protest against the introduction of GM crops in general, and against herbicide-tolerant plants in particular. One reason is the possibility that the use of GM-HT crops will stimulate the use of herbicides, another is the possible spread of the HT trait among non-crop plants.

The establishment of the HT trait in wild plant populations after seed and/or pollen dispersal is a valid argument against the use of GM-HT crop plants. There are fears that so-called superweeds could come into existence that would change the ecological balance in local niches. Wind, bees and other insects can carry pollen to another plant. Wind, birds, animals, humans and machines can carry seeds to new places where they can sprout and their pollen migrate. Problems include crop-related weeds that became herbicide-tolerant after out-crossing with transgenic plants. Cross-pollination may indeed occur with related species depending on the pollination system in the crop plant and the availability of target plant lines. While all of these arguments are reasonable, it must be realised that the trait will most likely be lost if there is no selective pressure to maintain the trait, i.e. if no herbicide is applied (which is unlikely in natural areas).

A relevant example is from Canada where three different varieties of GM rapeseed had been grown, each tolerant to a different herbicide (Roundup, Liberty, or Pursuit). After a few growing seasons rapeseed plants were observed to grow with tolerance to all three herbicides. Since these plants had formed on the fields without deliberate human planning or intervention, the plants could still arise because of relative unrestrained use of different types of herbicides. Farmers may therefore need to adjust their current practices in order to avoid this from happening. Among suggested recommendations were crop separation, crop rotation, a regulated use of herbicide formulations and improved communication with neighbouring farmers to adjust growth plans. The observed problem could easily be countered, however, by using a fourth herbicide to eradicate the triple HT rapeseed varieties. The Canadian superweeds were indeed removed with 2,4 D, a common herbicide that has been in use since 1946. This herbicide is also routinely used to eradicate naturally occurring herbicide-tolerant weeds, (80 reported types in the USA and Canada, and in 32 types in Australia).

An advanced possibility to prevent geneflow is to place transgenes in plant cell plastids. Plastids, like the plant chloroplast, are cell organelles that carry a set of genes that behave independently of the genetic material in the plant cell nucleus. Since plastids are present only in egg cells and not in pollen, the transgene could, in theory, only inherit through seeds produced in the maternal plant. It must be noted that this strategy does not offer a complete solution, since seed dispersal is still a problem and only a limited number of transgenes will function properly in plastids.

There are notable benefits that are associated with the use of HT transgenic plants. At the same time there are distinct problems that need to be addressed. Many of the controversies are not specific for the HT trait, and others concern the use of herbicides, a practice which is not specific for GM crops. Risk benefit analyses and up-to-date scientific data can help to decide on new directions to take in agriculture. New developments and opportunities must be explored if deemed appropriate, and should not be ignored as a result of pessimistic expectations.

2.1.9
Information Sources

Brooke G (2003) Roundup Ready soybeans in Romania, farm level impact. http://www.bio-portfolio.com/pdf/FarmlevelimpactRRsoybeansRomaniafinalreport.pdf

Commission of the European Communities (2002) Economic impacts of genetically modified crops on the agri-food sector, working document rev2. Directorate-General for Agriculture, Jambes, Belgium

Devine M, Duke SO, Fedtke C (1993) Physiology of herbicide action. Prentice Hall, Englewood Cliffs, NJ, pp 251–294

James C (2002) Global review of commercialised transgenic crops: 2002. ISAAA Briefs 27

OECD (1999) Series on harmonization of regulatory oversight in biotechnology, no 10. Consensus document on general information concerning the genes and their enzymes that confer tolerance to glyphosate herbicide. OECD, Paris

Orson J (2002) Gene stacking in herbicide-tolerant oilseed rape: lessons. English Nature. http://www.checkbiotech.org/pdf/englishnaturegenestacking.pdf

Wahlberg S (2000) Transgenic herbicide tolerance and male sterility in plant breeding. http://www.agsci.kvl.dk/breed/SBA/NordicBaltic/abstract3.htm

2.2
Insect Resistance in Crop Plants

Gert E. de Vries

The development of insect-resistant crops seems an ideal contribution to sustainable agriculture and could have benefits such as savings in resources devoted to scouting for pest insects, reduced applications of broad-spectrum insecticides, increased yields and protection against certain fungal plant pathogens. Over thirty different crops have been genetically modified to produce the *Bacillus thuringiensis* (Bt) delta-endotoxin throughout their plant structure. Any pest that feeds on any part of these plants will be exposed to this Bt protein, and those susceptible to the toxin will be killed. Balanced against potential benefits of transgene Bt usage are possible drawbacks: loss of Bt-mediated control due to resistance among populations of the target pest, exchange of genetic material between the transgenic crop and related plant species and the impact of Bt-crops on non-target species, possibly including humans by means of GM foods.

2.2.1
Introduction

In the next 50 years, the number of people living in the world's poorer countries will likely increase from 5 billion to 8 billion. This will require a 50% increase in agricultural products. Many technological developments in the past have improved crop yields. Without pesticides, 70% of the world food crop could be lost and even with current pesticide use, more than 40% is destroyed by insects and fungal damage. Almost a third of the world's cereal crop is lost to insect pests, mainly the larvae of moths and butterflies. Seven percent of the annual global maize harvest never reaches the market because of damage by the European corn borer. Insects are gradually evolving resistance to conventional pesticides, which also cause environmental damage and can contaminate food products. Insect damage may also lead to subsequent infection of crops by fungi, several of which are potent producers of carcinogenic mycotoxins.

The development of insect-resistant crops seems an ideal contribution and could have significant health benefits, savings on input and increased yields. A side effect is a better protection against certain fungal plant pathogens that make use of the wounds left by insects as starting points of infection. Possible drawbacks are the build-up of resistance among populations of the target pest, exchange of genetic material between the (transgenic) crop and related plant species, and the impact of the crop's properties on non-target organisms. In this section these issues will be discussed in more detail.

2.2.2
Management of Insect Pests

2.2.2.1
Agrochemicals

The use of agrochemicals was initially hailed as a great success during the green revolution, but soon turned out to have serious negative effects on the environment and human health. Manufacturers of many first generation pesticides, such as DDT, have now moved on to more readily degradable and safer products. Nowadays the regulation of agrochemical usage, choice of application, and measurement of residues on products are strictly controlled and are based on an extensive range of scientific data. Often it is not realised that toxicology studies of synthetic chemicals must be viewed in the context of the range of natural chemicals, which make up the vast bulk of substances to which humans are exposed. Natural and synthetic chemicals are similar in their toxicology and at the low doses of most human exposures, where cell-killing does not occur, hazards may be much lower than are commonly assumed.

The decades of use and development of pesticides to protect crops in order to increase yield nevertheless have shown that pesticides cannot be used indiscriminately. Major concerns like the persistence of pesticide residues in foodstuffs, development of resistance among pests and harm to non-target organisms must be addressed. Clever solutions may come from better knowledge of insect behaviour and physiology. A new generation of hormone-disrupting pesticides may, for instance, be developed to disrupt the life cycle of certain insects, preventing them from reaching their normal adult form. Because certain components of the insect hormone system may be very specific and have no equivalents in vertebrates, the chemicals potentially have a narrow spectrum and, most likely, are harmless to animals and humans.

2.2.2.2
Bt Spray

Bacillus thuringiensis (Bt) is a common soil microorganism that produces spores for its survival containing a protein crystal, the Bt-endotoxin. The protein has insecticidal properties affecting a selective range of insect orders, it is not toxic to mammals and shows fewer environmental effects than many synthetic insecticides. Bt spore formulations are regarded as an environmentally friendly approach because of their target specificity and decomposition to non-toxic compounds. For this reason organic farmers have embraced this technology, which is one of their few means of insect control. Over 30 different crystals (serotypes) have been identified among over 800 *B.thuringiensis* strain isolates, each with a characteristic range of effectiveness.

In order for the Bt toxin to be effective, the insect must ingest some of the Bt formulation. The crystal will be dissolved in the alkaline insect gut and activated by digestive enzymes. The endotoxin subsequently binds to the cells which are lining the midgut membrane and create pores that will disturb the ion balance in the gut. The insect stops feeding and starves to death. In the case that live Bt spore formulations are used and the insect is not susceptible to the direct action of the endotoxin, the insect may perish after the bacterium starts vegetative growth inside the gut. After the gut membrane is broken, a body-wide infection eventually kills the insect, the bacterium reproduces and makes more spores.

Recombinant techniques have also been used to introduce Bt genes into another bacterium, *Pseudomonas fluorescens*, where the toxin is produced in live bacteria. After killing the Pseudomonas cells the formulation is more resistant to degradation by UV and lacks bacterial spores or secondary toxins. The product is rejected by some organic review committees on the grounds that recombinant DNA technology was used during the production process.

2.2.2.3
Integrated Pest Management

Integrated pest management (IPM) is an ecosystem-based strategy that focuses on long-term prevention of pests or their damage through a combination of techniques such as biological control, habitat manipulation, modification of cultural practices, and use of resistant varieties. Pesticides are used only after monitoring indicates they are needed according to established guidelines, and treatments are made with the goal of removing only the target organism. Pest control materials are selected and applied in a manner that minimises risks to human health and the environment.

Insect damage to crops can for instance be reduced by the deployment of the natural enemies of target pests. Strains of the fungus *Beauveria bassiana* have been isolated with increased effectiveness at controlling caterpillar pests. The fungus *Metarhizium* is being developed for locust control. The formulations used to apply the fungal spores by aerial spraying are critical because the fungi need to remain alive long enough to come in contact with the insects.

Another example are species of *Trichogramma* wasps that attack the eggs of over 200 species of moths and butterflies, thus harmless to people, animals, and plants. The tiny wasps (0.5 mm) prevent crop damage because they kill their insect hosts before these can cause plant damage. Each female *Trichogramma* wasp parasitises about 100 eggs and may also destroy additional eggs by host feeding. The short life cycle of 8–10 days allows the wasp population to increase rapidly. Although *Trichogramma* occur naturally throughout the United States, they usually do not occur in high enough numbers to be effective at suppressing pest populations.

In some cases, delivery of the insect predators or the activation of a defense system can be left to natural processes. Physical damage of certain plants causes the release of the volatile distress signal methyl jasmonate, which attracts insect predators. Also plants are known to produce chemical compounds, peptide hormones or volatile terpenoids in response to bites by insect pests. These compounds are transported to the other parts of the plant, and serve as chemical signals repelling the pest or attracting natural predatory organisms and preventing further damage to the plant by the insects. It is feasible that some of these mechanisms may be stimulated and used in IPM.

2.2.2.4
Insect Resistance of Plants

A number of *in planta* strategies have been developed to combat insect damage in crop plants and are the subject of research in an effort to control insect pests. A very effective method to combat lepidopteran larvae has been to express the Bt protein in (parts of) crop plants, but a range of other approaches are worth mentioning as well.

2.2.2.4.1
Non-Bt Insect Resistance of Plants

Many wild and domesticated plants produce secondary compounds that effectively reduce or deter herbivory by insects and animals. Cyanogenic glycosides belong to one of the more toxic classes of herbivore deterrent compounds. By the introduction of the genes for the entire synthetic pathway of dhurrin from *Sorghum bicolor* to *Arabidopsis thaliana* it was demonstrated that the presence of the tyrosine-derived cyanogenic glucoside confers resistance to the flea beetle *Phyllotreta nemorum*.

Some plants are known to produce peptide hormones or volatile terpenoids in response to insect damage. These compounds are transported to the other parts of the plant, and serve as chemical signals repelling the pest and preventing further damage to the plant by the insects.

Trichomes, also known as leaf hairs, are specialised epidermal cells present in most plants that naturally excrete a host of lipid-based toxins. These serve to discourage predation by a wide range of herbivores, including microbes, insects, and large grazing mammals. Modification of the monoterpene biosynthetic pathways are promising ways to combat colonisation by aphids or damage by other insects, ants and animals.

A range of other, even more specialised strategies are being tested to defer or kill insects. These approaches are not aimed at a generalised use, but would only serve to safeguard specific plants from a particular type of insect attack.

2.2.2.4.2
Bt-Mediated Insect Resistance in Plants

Molecular genetic research has resulted in the development of crop plants capable of producing the efficient insecticidal proteins, derived from the bacterium *B. thuringiensis* (Bt), to replace the conventional spraying of broad acting pesticides. The *cry* genes code for a class of insecticidal proteins, bacterial delta-endotoxins, and are produced as crystals in *B. thuringiensis* (Bt-endotoxin) spores. The *cry* genes were chosen as a source of insecticide since live Bt biopesticide preparations are also effective when used to combat lepidopteran larvae feeding on crops. There are several strains of Bt, each with differing Cry proteins. There are over 100 patented Bt cry toxin genes including *cry1Ab*, *cry1Ac*, and *cry9C*, but most of the Bt-maize hybrids produce only the Cry1Ab protein.

Over 20 transgenic crops with the Bt insecticide have been field-tested in the United States, and three (maize, cotton, and potato) have been widely planted. The major pest controlled by Bt-maize is the European corn borer and Bt-cotton is used to control the cotton and pink bollworm and the tobacco budworm. Other important insect pests where there is some control include the cabbage looper, saltmarsh caterpillar, cotton leaf perforator, corn earworm, southwestern corn borer, and other stalk boring insects. The Bt-endotoxin is regarded as an environmentally friendly insecticide because of its target specificity and its decomposition to non-toxic compounds.

2.2.3
Benefits and Risks

During the six-year period 1996–2001, herbicide tolerance had consistently been the dominant trait in GM crops, with insect resistance being second. In 2001 Bt-crops occupied 7.8 million ha, equivalent to 15% of the global transgenic area. Stacked genes for herbicide tolerance and insect resistance were employed in both cotton and maize and these crops were planted on 8% of the transgenic area. It is expected that the trend for stacked genes will continue to gain an increasing share of the global transgenic crop market.

2.2.3.1
Economic, Environmental and Food Safety Benefits

The premium paid by farmers for Bt-crop seeds will likely only be returned in years when target insect pest infestations are moderate to heavy. Farmers will thus be inclined to pay the fee when they anticipate that benefits will exceed the premium. Actual benefits at the end of the growing season can be less when growers face unexpected insect pests which are not controlled by Bt-endotoxins and the savings on reduced chemical use never materialises. An investment in Bt-crops is therefore an economic risk and the principle of insurance is a good analogy. The variability in target pest infestations and market prices raises for example concerns about fluctuations in yearly economic benefits of Bt-maize.

Environmental benefits refer to the indirect positive environmental or human health effects that benefit society as a whole, but which are not captured in direct costs or returns. Examples of environmental benefits may include less worker exposure to chemical insecticides, less ground and surface water contamination, and less impacts on non-target wildlife. Because Bt toxins are highly specific to target insects, Bt-crops, if used cautiously, offer ecological benefits over conventional broad-spectrum insecticides such as the pyrethroids. The estimation of the value of environmental and human health benefits has not reached consensus. The range of benefit projections are often too wide for meaningful interpretation.

Food safety benefits arise from the fact that pathogenic fungi may gain access into crop plants when damaged by insect pests, leading to rot of plant tissues. The fungi can release mycotoxins, chemicals toxic to both animals and humans. Mycotoxins include aflatoxins (produced by fungi in the genus *Aspergillus*) and fumonisins (produced by several species of *Fusarium* fungi). Both aflatoxins and fumonisins can be fatal to livestock and are probable human carcinogens. Several studies have supported the utility of Bt hybrids for management of *Fusarium* and *Aspergillus* ear rots and stalk rots of maize, although not all Bt transgenic plants are equally effective against the full spectrum of insects that can contribute to kernel damage and subsequent mycotoxin contamination.

2.2.3.2
Risks

2.2.3.2.1
Environmental GM-Bt Effects

The US Environmental Protection Agency has evaluated studies of potential effects on a wide variety of non-target organisms that might be exposed to the Bt-endotoxins, e.g. birds, fish, honeybees, ladybugs, parasitic wasps, lacewings, springtails, aquatic invertebrates and earthworms. Such non-target organisms are important to a healthy ecosystem, especially the predatory, parasitic, and pollinating insects. These risk assessments demonstrated that Bt proteins expressed in transgenic plants do not exhibit detrimental effects to non-target organisms in populations exposed to the levels of endotoxin found in plant tissue.

2.2.3.2.2
Residues, Root Exudates/Breakdown

What about less visible effects? Does the Bt protein enter the soil system from transgene plants (leakage from roots, decaying plants), would it persist and would any soil organisms, including micro-organisms, be affected by its presence?

Plant roots only secrete a limited number of specialised proteins, more specifically those that are provided with a specific coding signal for export. The Bt proteins contain no such secretion sequence code, and therefore leakage of the protein from live roots will be minimal. Decomposing plants therefore will be the main source of Bt protein in soils, where it could bind to soil particles and form a route of exposure to soil organisms. Persistence or half-life may be expected to vary significantly depending on soil conditions and is reported to vary from 1.6–22 days, but has been measured to be as long as 46 days.

Organisms that pass soil through their digestive systems, such as earthworms, could be exposed to significant levels of Bt protein. Organisms at higher trophic levels that feed on soil feeders may, secondarily, be exposed to the toxin. High dose and continuous feeding studies on Collembola (primitive wingless insects that live off the fungi that decompose organic matter) and soil mites did not indicate likely adverse effects on such non-lepidopteran species in the field. It is clear that it will be impossible to test every individual species of soil organism. While no significant impact of Bt protein in soils on higher soil organisms is known today, risks seem theoretically plausible.

Small transient changes in microbial populations have been associated with some Bt transgenic plant material, but many other factors have been shown to cause more significant changes in microbial populations in agricultural and other soils. Without better information regarding the range of what constitutes natural microbial communities or microbial communities in current agro-ecosystems, and the consequences of such changes, it is not possible to assign a significance to apparently minor changes in microbial populations if they occur.

2.2.3.2.3
Bt Pollen and Non-Target Organisms

Dispersal of pollen that also may, but not necessarily, contain the Bt protein from transgene crops has been intensely debated as a threat to non-target lepidopteran insects, with the Monarch butterfly being the most prominent victim. Roughly 50% of the Monarchs in the USA may pass through the Corn Belt each year, 10% of the potential summer Monarch breeding range is estimated to include maize fields. After laboratory force feeding studies showed the toxicity of Bt protein for the Monarch, a range of studies have been performed to estimate the probability of Monarchs or other butterfly species encountering toxic levels of pollen inside, at the near edge or at any distance away from a field of Bt-maize. It was concluded that survival of butterfly populations is likely to be strongly influenced by factors other than Bt, such as loss of habitat and climatic conditions.

Considering the gains that are obviously achieved in the level of survival of populations of Monarch butterflies and other insects by eliminating a large proportion of the pesticides applied to maize, cotton and potato fields, some authors are predicting that the widespread cultivation of Bt-crops may have huge benefits for Monarch butterfly survival. However, indirect impacts of Bt-maize on natural enemies of crop pests may occur as well. Predators, parasites and pathogens might decline as pest populations decline, and minor pests may become more predominant when applications of foliar insecticides are reduced.

2.2.3.2.4
Geneflow and Horizontal Gene Transfer

Wild plant species related to the main transgenic Bt-crops used (maize, potato and cotton) cannot be pollinated by GM varieties due to differences in chromosome number, flowering time and habitat. Therefore geneflow from Bt-crops, so far, poses no significant environmental risk. However, geneflow can pose an economic, agricultural risk when GM crops contaminate non-GM crops such as may be the case with maize.

DNA from crop plants has been shown to persist in soils, protected from degradation by binding to clay or organic components, at detectable levels for at least several months to several years. Uptake and integration of such DNA into micro-organisms, so-called horizontal transfer, can only been accomplished at low frequencies and under optimised laboratory conditions. Horizontal gene transfer therefore is not regarded as a serious or significant risk.

2.2.3.2.5
Insect Resistance to Bt-endotoxin

The observation that over 400 insect species have become resistant to at least one insecticide is an indication of the impressive genetic ability of arthropods to evolve resistant strains. During the development of Bt plants, scientists sought to have high expression of the toxin throughout the plant and throughout the rowing season, in order to avoid the appearance of resistance in insects. The practice of using Bt spore formulations expose insects to Bt only intermittently and at varying doses and may have resulted in the selection of insect variants with levels of low resistance to Bt.

A key element of preserving the long-term effectiveness of the Bt technology rests on delaying the development of insect pest resistance through the use of resistance management plans. This is currently based on two complementary principles: high dose and refugia. High dose refers to the fact that Bt-crops should produce sufficient levels of Bt toxin to kill all susceptible larvae, even those carrying one copy of a resistance gene. Refugia refers to planting a portion of each field with non-Bt-crops to allow for interbreeding between insects which may have acquired resistance and insects that are still susceptible to Bt. New technologies to prevent the occurrence of resistance include Bt expression modes for specified periods of time (effectively creating larger refuges but minimising crop loss) and expression of dissimilar toxin genes in the same plant. A third option is the use of transgenic insects to breed large numbers of sterile males and control unwanted insects by their release.

2.2.3.2.6
Human Food Safety

While the Bt crystal protein is a toxin for certain insects, it is certainly safe for human consumption because the crystal Bt protein is digested like any other food protein and is denatured (i.e. the specific activity is destroyed) by the acidity of the human stomach. Therefore it was no surprise that when feeding exorbitant high amounts of Bt proteins to rats, the animals showed no negative effects. Another issue raised for food safety relates to the possible allergenic character of certain Bt constructs and this matter received great attention of the press when Starlink maize, containing the Cry9C protein and regulated in the USA for feed purposes only, turned up in Taco Bell's products (a fast food restaurant chain).

Common food allergens tend to be resistant to degradation by heat, acid, and proteases; the proteins may be glycosylated and are generally present at high concentrations in food components. Bt proteins that turn up in food products do not meet these specifications since the protein shows no coding similarities with known allergens. Decades of widespread use of *Bacillus thuringiensis* as a pesticide (Bt formulations have been registered since 1961) have not resulted in confirmed reports of immediate or delayed allergic reactions to the delta-endo-toxin itself despite significant oral, dermal and inhalation exposure to the microbial product.

Specific concerns have been expressed about Cry9C, the Taco Bell Bt toxin, which possesses some heat- and digestion-resistance but no conclusive allergenic activity could be demonstrated in studies dedicated to this specific protein.

2.2.4
Information Sources

AGCare fact sheet. Bt and the Monarch butterfly. http://www.agcare.org/monarch1.pdf

Bt Corn is not a threat to Monarch. http://www.ars.usda.gov/is/AR/archive/feb02/corn0202.htm

Example biopesticide fact sheet (006458) *Bacillus thuringiensis* Cry1Ab delta-endotoxin and the genetic material necessary for its production (plasmid vector pCIB4431) in corn (event 176). http://www.epa.gov/oppbppd1/biopesticides/pips/bt_brad2/3-ecological.pdf

Industry insect resistance management plan for Bt field corn. http://www.ncga.com/biotechnology/insectMgmtPlan/importance_bt.htm and http://whybiotech.com

Issues pertaining to the Bt plant pesticides risk and benefit assessments by the U.S. Environmental Protection Agency with links to background documents. http://www.foodsafetynetwork.ca and http://www.biotech-info.net/bt-transgenics.html

US EPA prepared a 283-page detailed summary of data about risks and benefits of insect-protected Bt-crops:Bt biopesticides registration action document. Report on the economics of Bt transgenic crop usage. http://www.bio.org/food&ag/bioins01.html

2.3
Virus Resistance

Ervin Balazs, Paul Pechan

Before the advent of genetic engineering, temporary resistance to plant viruses was based predominantly on traditional plant breeding efforts. This approach was combined with various crop rotation strategies and spraying against organisms that may carry viruses. Genetic engineering has made it possible to isolate genetic material from viruses themselves and transfer this material into plants to induce resistance against virus infections. The strategy is based on knowledge accumulated in traditional plant pathology. It is called "cross protection", where a mildly infectious virus strain could produce protection against a related highly infectious strain. The idea of cross protection is similar to people getting flu shots to become less susceptible to flu epidemics. It was shown in 1980s that even parts of a virus could offer plants this cross protection. When the gene responsible for the synthesis of a protein that coats the viral genetic material was introduced into tobacco, it led to viral protection as effective as the classical cross protection induced by an entire mild strain of the given virus. The initial cross protection success was followed by examples with other host plants and other viruses. The coat protein gene when introduced into host plants is called transgene. Today, a number of genetically modified virus-resistant crops, based on the coat protein cross protection, are in an advanced stage of commercialisation. In China alone, hundreds of hectares are planted with genetically modified virus-resistant tobacco and tomatoes. To date, there are no indications that would suggest that this technology has had a negative impact on the environment. However, scientists have raised several questions that need to be answered before genetically altered virus-resistant plants can be introduced into the environment on a large scale. These concerns relate to:

1. The potential that the coat protein may help the spread of the infecting virus via heteroencapsidation
2. Recombination between the genetic material of an infecting virus and the transgene
3. Synergistic effect of the transgene and the virus in the disease development
4. Spreading of the virus resistance transgene by cross pollination to wild relatives

This section summarises information concerning the biosafety considerations of crop plants that are made virus-resistant through genetic modification, with special reference to the coat protein mediated protection.

2.3.1
Introduction

Plant viruses can cause substantial economic losses when genetically homogenous plant varieties are grown in large numbers and in one location (the so-called monocultures). Viruses do not usually cause significant losses in wild plant populations because the plants are genetically more diverse and they tend not to grow in high densities in one area. The well-documented losses caused by viruses in perennial crops such as trees are known and often cited in the literature. Cacao trees in Ghana in the 1930s were dramatically affected by the cacao swollen shoot virus. The virus, transmitted plant to plant by mealy bug, led to an epidemic that was eradicated after cutting out more than 160 million trees by 1977. In Europe, a similar devastating virus called "Sharka", or plum pox virus, seriously affects fruit plants, such as apricot, plum, peach and several ornamental Prunus species. The virus spread all over Europe in the last century from Bulgaria where it was first reported in 1915. When a beet necrotic vein virus, transmitted by a fungus, spread out in Europe in the last decades, it caused serious losses to the sugar beet industry because of reduced sugar content of the infected beets. Enormous losses are also due to an aphid transmitted citrus tristeza virus in citrus plantations. Millions of trees were killed and had to be destroyed due to this virus disease: for example in Sao Paolo State of Brazil 6 million, in Argentina 10 million, in California 3 million and in Spain 4 million trees. In potato, different viruses can accumulate during successive vegetative propagation. This led to the disappearance of several high quality potato cultivars from today's agricultural practice. Early in the last century, leafhopper-transmitted beet curly top virus almost eradicated the sugar beet industry in California and neighbouring states. One could list many more examples to demonstrate the economical and ecological impact of viruses on commercial crops.

Newly emerging virus diseases are recorded when their presence causes significant economic losses or the epidemic seriously affects the cultivation of certain crops. As plant viruses may cause serious losses, new crop varieties must be developed either by conventional plant breeding or by the use of genetic modification to counteract their effects.

2.3.2
Characteristics of Plant Viruses

2.3.2.1
How are Plant Viruses Named?

Plant viruses had been named after the plant species from which they were first isolated. The names also contain the characteristics of disease symptoms on the plants from where they were first isolated. For example, the tobacco mosaic virus was first isolated from a tobacco showing mosaic spots on the leaves. Viruses are now named based on the structure and organisation of their genetic material.

2.3.2.2
Evolution of Viruses Through Recombination

The likelihood that a new viral strain will be viable and survive depends on how well it can compete with other viruses, especially on its capability to copy its genetic material in the presence of the virus it originated from (called parental virus), and its successful spread within a plant. Based on the large database of genetic information on plant viruses already collected, it is now possible to analyse the relationship of the different virus species. It has thus been shown that certain viral genes probably arose by recombination events, in the process creating new viral strains that compete with the parental viruses. Recombination plays an important part in virus evolution. Currently, it is not possible to determine how these events occurred as they likely occur over a much longer time frame than we have been able to observe and measure experimentally in contained greenhouse and field trials. In addition, the likelihood of recombination under natural conditions and in the absence of selection pressure has not been determined. This means that obtaining reliable data from direct observations is very difficult. Such information is important because it is needed as baseline data to investigate the possible effects of virus-resistant plants on virus evolution. There is, however, some evidence that the genetic makeup of viruses is quite stable since, for example, the tobacco mosaic virus has been stable for over 100 years.

2.3.2.3
Structure and Transmission of Viruses

2.3.2.3.1
Virus Structure

Plant viruses have a relatively simple structure composed of genetic material in the form of either ribonucleic or deoxyribonucleic acids (RNA or DNA), surrounded by a protein shell, called capsid. Some capsids may also contain lipids and/or carbohydrates. The viruses are classified according to the type of their genetic material (ie whether they contain RNA or DNA). More than 95% of plant viruses contain RNA as their genetic material, either in single-stranded or double-stranded form. Very few plant viruses contain DNA. The genetic material of the virus usually comprises of several RNA-based genes coding for a range of essential functions. These include, for example, an enzyme that helps in copying the genetic material of the virus (called replication), a protein which coats and protects the genetic material (called capsid protein), a protein which is responsible for the movement of the virus from cell to cell and several other components that have specific roles in the replication cycle and in the infectivity of the virus. The host plants have, in addition, specific genes that influence the host-virus interaction.

2.3.2.3.2
Virus Transmission

Viruses are transmitted from plant to plant either mechanically or with the help of carrier organisms such as insects. The common mechanism for viral infection is that these carrier organisms injure a plant, providing an entry point for the virus into a plant cell. Plant cells have a strong wall that normally prevents viral infections. That is why viruses are usually carried plant to plant by organisms that disrupt the plant cell walls and thus make the virus entry into a plant possible. After entering a plant cell, the virus uncoats, i.e. its genetic material separates from the protective protein shell. The virus genetic material will then be replicated relying on the plant cell machinery. The last step of the virus replication is the so-called encapsidation when the free viral nucleic acid is covered with the protein shell.

The process is then repeated, leading to an ever-increasing number of virus particles within one cell. The virus then moves into other plant cells through channels connecting cells (called plasmodesmata) or in specific cases through the nutrient conduction tissue of the plant (called phloem). In this way, the virus can infect the whole plant and produce visible disease symptoms. Usually this is manifested through growth retardation, mosaic pattern on the leaf, ring spots, wilting and other disorders. In certain cases the virus infection leads to cell necrotisation or death, called lethal necrosis.

Viruses vary greatly in the range of species they are able to infect. Some viruses affect only certain plant species, while others are able to infect a wide range of plants. Viruses may be transmitted alone, or they may be transmitted together with other viruses or virus strains. When two virus strains are being replicated within the same cell, two different capsid proteins are produced. This can lead to a mixed coating of the virus genetic material, called transencapsidation. Such viral particles may or may not be sufficiently functional to allow transmission of the virus to another host plant. This mismatched protein shell cannot be maintained in the subsequent hosts as no changes in the genetic make up of the viruses are involved.

2.3.2.3.3
Synergy

Plant pathologists often describe that when two different viruses simultaneously infect plants, the symptoms can be more severe than when only one of the viruses

infects the plant. This phenomenon is called synergy. The result of such an infection is that the crop cannot be marketed. The mechanisms of the synergism
between two viruses is not well understood. Such a synergistic effect can complicate the design of viral resistance in crops.

2.3.3
Conventional Plant Virus Management

Besides plant breeding for virus resistance, several agronomic alternatives are
used to combat viral infections. These are, for example, spraying of viral carriers,
encouraging virus-free plant cultivation, application of strict phytosanitary practices and crop rotation. Some of these approaches are discussed below.

2.3.3.1
Resistance Breeding

One of the major efforts to combat virus infections is by breeding virus-resistant
crops. This approach can be used if sources of resistance genes are known. Plants
can be resistant to viral infections if they contain viral resistance genes. The
resistance is usually manifested by lack of an infection or, if an infection does
occur, through a hypersensitive reaction in an area where the infection occurred.
In traditional plant breeding, virus resistance genes could be obtained from
wild relatives of cultivated crops. These genes are, however, usually linked to
other genes that are undesirable in the new crop variety and co-introduced
with the resistance genes. To eliminate unwanted genes is a time-consuming
and very laborious plant breeding process. Even if this goal is achieved, the
resistance will eventually breakdown due to the fact that there are always new
viral strain populations that could become abundant after a new crop variety is
introduced.

2.3.3.2
Carrier Control

Conventional control of viral diseases in agriculture is based on soil fumigation
and chemical sprays for the elimination of virus carriers, such as aphids, mealy
bugs, leafhoppers, thrips, fungi and nematodes. Additional techniques include the
exclusion of contaminated materials and the planting of virus-free stocks or
seeds. The most effective way to protect crops against virus diseases is the use of
chemical sprays against the carriers responsible for the transmission of the virus.
However, extensive use of pesticides should be reduced as much as possible due
to possible damage to the environment and possible health risks.

2.3.3.3
Cross-Protection

A mildly infectious (attenuated) virus strain can induce protection of the host
plant against other aggressive viral infections. This is called cross protection.

There are several explanations for how cross protection may work. One of the possible mechanisms for cross protection is when a protein, from an attenuated virus, coats the infecting viral nucleic acids and prevents their replication. Recently, it was shown that preventing gene transcription in the aggressive virus (called gene silencing) is yet another reason for virus resistance in plants. This technology raises two major concerns. Firstly that the potential mutation of the mild strain could lead to the formation of an aggressive virus and secondly that the strain used for protection may induce severe diseases in other crop plants. Concerns about the use of cross protection appeared to be resolved by genetic engineering when it was shown that integrating and expressing a non-replicating virus coat gene in the host genome is able to induce the cross protection phenomenon.

2.3.4
Plant Virus Management by Genetic Modification

It is hoped that genetically modified crops will reduce the costs both to the environment and the farmer, for example by reducing the reliance on spraying of virus carriers and more quickly introducing new virus-resistant varieties onto the market. The main advance to date in this respect came through using the viral coat protein approach.

In 1986 an American research team developed virus-resistant tobacco and tomato plants by integrating into them the coat protein gene of tobacco mosaic virus. They obtained plants that were resistant to this virus. This initial success was followed with other examples using other plants and viruses. Several large scale field trials were carried out. The coat protein mediated protection technology is now used for commercial purposes. It is considered more safe than conventional cross protection because only a very small part of the virus genome is used to protect plants. This contrasts the use of the whole viral genome in conventional plant breeding efforts. Besides the coat protein mediated resistance strategy, several new approaches, still in an experimental phase, are being developed for protecting plants against viruses. These include the use of several other viral DNA and RNA sequences (for example satellites or replicase genes). Another possibility is to use the so-called plantibody technology. Plants do not have an immune system like that of animals in which specific antibody proteins are formed in response to an infection. But, expressing genes in plants responsible for producing antibodies in mice to a specific virus, made it possible to build an artificial immune-system.

The coat protein mediated protection is efficient and so far has proved to be safe. Nevertheless, scientists have raised several questions concerning the risks associated with the coat protein mediated approach. These concerns are discussed below.

2.3.5
Potential Risks of Using Genetic Engineering to Fight Plant Viral Infections

2.3.5.1
Hetero/Transencapsidation in Virus-Resistant GM Crops

The potential environmental impact of interactions between viruses and viral coat protein protected plants is generally expected not to be more serious than the impact that occurs in mixed infections in susceptible hosts. There are examples of interactions that naturally occur between plant viruses, such as occurring through transencapsidation, where coat proteins get interchanged between the viruses. Similar interactions may occur in transgenic plants. The approach is to use only non-functional coat protein when engineering virus resistance. It has been observed that even if the coat protein is produced in transgenic plants, limited or no virus replication was detected. Consequently, the possible amount of mismatched virus production is far less than in a susceptible, non transgenic host. Currently used genes, encoding non-functional proteins, could be efficiently used for cross protection against viral infections without increased danger of mixing coat proteins.

2.3.5.2
Recombination in Virus-Resistant GM Crops

Most virus-resistant transgenic plants in use today contain genes encoding coat protein of viruses that regularly infect the host plants and that induce the most devastating losses in crop production. In most of these plants, the virus gene used to induce sufficient protection against the target virus is usually expressed at much lower levels than in non GM susceptible plant where it is expressed together with all the other infecting viral genes that contribute to viral replication and spread within a plant. The general objections to using virus-derived protection are based on the possibility of recombination between the infecting virus genome and that of the RNA originating from the protein coat transgene that protects the GM plant. Indeed, this type of recombination has been observed under experimental conditions. In GM crops, the transgene giving the plant viral protection is constantly produced in the plant cell. When a virus infects the GM crop, the virus nucleic acid could recombine with the transgene. However, in the resistant plant, virus replication and associated protein synthesis is either not detectable, or very limited, while in a susceptible plant this could reach up to 10% of the total protein content of the plant. The recombination is thus significantly higher in the susceptible plant than in a resistant one due to the difference in replication level of the virus in those two types of plants. Moreover, the genetic sequences involved in the recombination events can be eliminated, leaving only the genetic part that offers viral protection. Sequences from engineered viruses are thus unlikely to pose a potential for generating novel recombinants at higher frequencies compared to mixed infections in nature.

   Although recombination is part of the virus evolution and although virus genetic makeup appears to be quite stable over short periods of time the effect of large scale use of transgenic plants on the evolution of plant viruses still needs to be determined.

2.3.5.3
Viral Synergy in GM Crops

Double infection can lead to the synergy phenomenon, which is often manifested by worsening of the symptoms. There are some experimental data suggesting that in certain cases a coat protein gene introduced into a plant or its product will be sufficient to induce this synergistic effect in the specific transgenic plant. Moreover, the use of coat protein mediated resistance might open the possibility for new types of interactions between different viruses. As the transgene encoding the coat protein can be expressed in all cells of the host plant, it is theoretically possible that a virus that normally limits its activity to only a small part of the plant might develop a synergistic interaction with the transgene product. This may lead to a new type of symptom (disease) or modification of movement of the virus within transgenic plants and transmission between plants.

2.3.5.4
Gene Escape

Geneflow between cultivated plants and their wild relatives has occurred since the beginning of plant domestication. In the case of cross pollinated plants, the transgene of viral origin can escape when present in a pollen that pollinates wild relatives of the transgenic crop. The question is what effect will the transgene have on the wild plant population. Even if a virus resistance gene escaped to wild relatives of the cultivated crop, it would not necessarily give that plant a selective advantage.

Although transgenic crops are unlikely to pose a greater problem to the environment than those obtained through conventional plant breeding, more information is needed on the impact of virus-resistant plants (transgenic and non-transgenic) on the natural ecosystem. This includes for example studies on selective advantages of virus-resistant plants and effects of viral synergies.

2.3.6
Information Sources

European Plant Protection Organisation. The site contains retrievable information about plant viruses. http://www.eppo.org
ICTVDB: The Universal Virus Database contains an overview of currently known viruses. http://www.ictvdb.bio2.edu
Matthews REF (1993) Plant virology. Academic, San Diego, CA
OECD consensus documents provides information on biosafety issues, including a series of document related GM plants. http://www.oecd.org

2.4
GM Crops in the USA and Europe:
What Farm Level Benefits Might be Expected?

Anne-Katrin Bock

Genetically modified (GM) crops have been adopted to a large extent in several countries during the last few years. In the EU, GM crops are still the focus of a controversial discussion, focusing on potential risks and uncertain benefits. There have been several surveys and model calculations, mainly for the USA, to obtain information on impacts on crop yields, pesticide use and change of agricultural practices due to application of GM crops. These studies indicate positive effects of herbicide-tolerant and insect-resistant GM crops on reduction of herbicide and insecticide use with potential environmental and economic benefits as well as a trend towards the more environmentally friendly no-tillage system. Comparable effects have also been calculated for EU agriculture. These results still need to be verified but should be taken into consideration in the discussion on GM crops. GM crops certainly are not the only means by which agriculture can be rendered more environmentally friendly, but they could contribute to that aim.

2.4.1
Introduction

Genetically modified (GM) crops were introduced to agriculture in 1996. Between then and 2002 the global area cultivated with GM crops has increased from 1.7 million ha to 58.7 million ha (James 2002). The use of GM crops is not distributed equally over the world but focuses on four countries, which in 2002 grew 99% of the global GM crop area: USA (66%), Argentina (23%), Canada (6%) and China (4%). In the EU only Spain and Germany are mentioned as growing GM crops in 2001 (James 2002a). Other European countries listed with small areas for 2002 are Rumania and Bulgaria. All in all 13 countries were growing GM crops in 2001; the number increasing in 2002 to 16 with India, Colombia and Honduras.

As can be seen in Fig. 2.1, soybean, maize, cotton and canola are the main GM crops cultivated. In 2002 soybean represented 62% of the global GM crop area, maize 21%, cotton 12% and canola 5%. Compared to the global crop specific area, the adoption rate for GM soybean is 51%, for GM cotton 20%, for GM canola 12% and for GM maize 9% (see Fig. 2.2). GM crops currently grown on a commercial basis mainly exhibit agronomic or so-called input traits. Herbicide tolerance represents the most frequent trait (75% of global area grown with GM crops), followed by insect resistance (17%) and combined traits for herbicide tolerance and insect resistance (8%).

Although it seems that the use of GM crops is expanding rapidly there are still ongoing controversial discussions about potential environmental and health risks (European Parliament and Council 2001) on the one hand and uncertain benefits of the technology on the other hand. In the EU since 1999 no new authorisation for commercial cultivation of GM crops has been granted. This was a reaction to consumer concerns in the EU and perceived shortcomings of the then

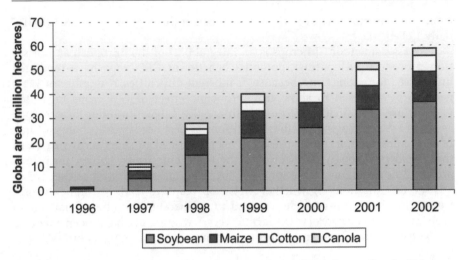

Fig. 2.1. Increase of global GM crop area from 1996 to 2002 and shares of main GM crops (source: James 2002)

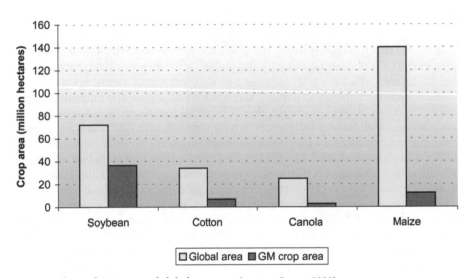

Fig. 2.2. Share of GM crops of global crop area (source: James 2002)

existing regulations. In October 2002 the revised directive on deliberate release of GMOs in the environment (2001/18/EC) went into force. New legislations are also in place for GMO traceability and labelling as well as food and feed. These are intended to improve transparency and efficiency in order to increase consumer confidence in GM products as well as in the process of authorisation. Based on these developments, the (re)start of the authorisation process is discussed. Several new notifications for placing GMOs on the market have been sent from some member states to the Commission at the beginning of 2003 (ENDS daily 19 Feb 2003).

Risk assessment of GM crops is an integral part of the European authorisation procedure. In contrast, potential benefits of GM crops have remained unassessed for a long time, because of inherent difficulties. A number of years need to be analysed to show any agronomic effect of GM crops regarding yield or use of pesticides, because both are also subject to other factors, such as year to year changing weather conditions or pest occurrence, change of mix of pesticides used, fertiliser use etc. This might lead to short-term variations that could be wrongly connected to the transgenic traits. Longer term data are until now only available for the USA. Thus nearly all studies that try to answer the questions of whether GM crop planting leads to higher yields, less herbicide and insecticide use, and to a different, more beneficial way of crop management concern US farms. Very few attempts have been made to investigate actual or potential agronomic and economic impacts of GM crops for the EU.

This section summarises the recent reports and publications on the situation in the USA and Europe to give an overview on available information.

2.4.2
Main Traits of Current Commercially Planted GM Crops

The main GM crop trait is herbicide tolerance with 75% of the global GM crop area in 2002. Glyphosate (Roundup) and glufosinate (Liberty or Basta) are the two key herbicides, which are active against a broad spectrum of weeds, against which tolerance has been introduced into GM crops. Glyphosate acts by inhibiting a specific enzyme in plant chloroplasts that is essential for the production of certain amino acids and derived substances. Glyphosate-tolerant GM plants received a bacterial gene coding for a glyphosate-tolerant version of the enzyme. Glufosinate inhibits an enzyme which is involved in the general nitrogen metabolism of plants, resulting in a toxic concentrations of ammonium in plant cells. Glufosinate-tolerant GM plants were supplied with a bacterial gene that inactivates the herbicide.

Insect resistance is conferred to plants by the introduction of a gene from the soil bacterium *Bacillus thuringiensis* (Bt). So-called Bt-plants express the protein encoded by the gene which, when ingested by lepidopteran insects, acts as a toxin and binds to the gut membranes causing leakage. The toxin acts selectively against lepidopteran species. The Bt toxin is very well known and is also in use as a spray, for example in organic agriculture.

Virus resistance of a plant can be achieved by introducing a selected viral gene, coding e.g. for a viral coat protein. The expression of the coat protein does not hurt the plant and renders it resistant to that virus.

2.4.3
GM Crop Performance in the USA

Gianessi et al. published in June 2002 the results of 40 US case studies on GM crops designed to improve pest and weed management. They analysed available

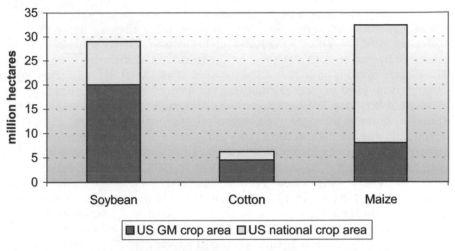

Fig. 2.3. US area for conventional and GM crops for soybean, cotton and maize for 2001

data for the eight GM cultivars planted on a commercial scale in the USA in 2001: Bt-corn and cotton, herbicide-tolerant canola, corn, cotton and soybean, and virus-resistant papaya and squash.

Based on data from the US states with the main cultivation areas, changes in production cost, crop yield, and pesticide use have been analysed. The results of six of the case studies are summarised and discussed here.

Figure 2.3 gives an overview of the adoption rate of GM soybean (herbicide-tolerant), GM cotton (herbicide-tolerant and insect-resistant) and GM maize (mainly insect-resistant). GM soybean and cotton have been adopted quite significantly by US farmers, in contrast to GM maize, which represents up to now "only" 20% of the national maize area.

2.4.3.1
Herbicide-Tolerant Soybean

GM soybean is the main GM crop cultivated world-wide. In 2002 GM soybean represented 51% of the total soybean area of 72 million ha. In the USA the only GM soybean planted is herbicide-tolerant (HT) soybean, tolerant against the herbicide glyphosate (Roundup Ready varieties).

Soybean is the second largest crop after maize in the USA. Some 29 million ha of soybean were cultivated in 2002, 69% being GM HT soybean (20 million ha). In Argentina nearly 98% of the national soybean area of 11.4 million ha (2002) is planted with GM soybeans. The study undertaken by Gianessi et al. (2002) included the 31 main soybean growing states in the USA.

The question of yield gain or loss has been discussed in several publications (Benbrook 2001, Carpenter 2001, Fulponi 2000). It seems that during 1998 and 1999, based on a variety of field trials in eight northern US states, HT soybean on average yielded 4% or 3% less than conventional soybean. As no isogenic varieties have been compared, the reason for the lower yield might be the different agronomic background of the varieties rather than a yield reduction (so-called yield drag) due to the Roundup Ready trait. Variety trials from 2000, presented by Benbrook (2001), show reduced yield of GM soybeans of up to 11%. The picture remains ambiguous and Gianessi et al. (2002) did not include any differences in yield in their calculations, but focused on herbicide application. Planting of Roundup Ready soybeans was compared with a plant management aimed at obtaining the same weed control result as with glyphosate.

Natural weed infestations can cause yield losses of 50–90% if untreated. Weeds compete with soybean for water, nutrition and sunlight. Additionally they decrease the efficiency of harvesting equipment. In 1995, 23% of the soybean area was treated with a combination of four or more active pesticide ingredients; 28% were treated with three different ingredients, 35% with two and 12% with one. Until today more than 150 seed companies offer more than 1,000 Roundup-Ready soybean varieties in the USA. HT soybeans deliver results comparable to weed-free situations. Herbicide use pattern changed accordingly. The glyphosate usage increased to 75% of the national soybean area treated, because of its effectiveness for a broad range of weed species. Another effect is changed tillage practice with a clear trend towards fewer tillage passes and conservation tillage (average reduction about 1.8 tillage/ha).

Total costs of the Roundup-Ready technology are estimated at $38.3/ha. Application of glyphosate is estimated at 1.06 kg/ha active ingredient. Herbicide costs are assumed at $475 million. Additionally a technology fee of $14.81/ha or $300 million has to be taken into account (see Table 2.4).

To estimate the impact of HT soybean use, herbicide replacement scenarios were analysed based on surveys in different states. To reach the same effect as Roundup Ready without the need for additional cultivation, alternative programmes on average use three different compounds and result in costs of about $74–99/ha (calculated average $88.2/ha).

Based on these figures about $1 billion/year can be saved by growing Roundup Ready (RR) soybean with a reduction of 13,000 t of herbicide active ingredients applied. This translates into a cost decrease of $50/ha.

Other reports present slightly different figures for herbicide use. According to Carpenter et al. (2002), in 1995, one year before GM soybean introduction, 1.12 kg/ha were used for soybeans and no significant decrease was reported until

Table 2.4. Costs of GM soybean cultivation compared to conventional soybean

	GM soybean	Conventional soybean
Herbicide use and cost	1.06 kg/ha, 21,500 t/year, $475 million	1.71 kg/ha, 34,506 t/year, $1,786 million
Technology fee	$300 million, ($14.81/ha)	—
Total cost	$775 million or $38.31/ha	$1786 million or $88.23/ha
Difference	$50/ha or $1,011 million: 0.65 kg/ha herbicide active ingredient or 13,000 t herbicide active ingredient	

2000. Others cite an increase of 3% as a result of adoption of GM soybean. GM soybean farmers generally use fewer different herbicides (on average 1.4 active ingredients per ha) while conventional soybean farmers use 2.8 active ingredients per ha. This is supported by the increase of glyphosate use since introduction of GM soybeans and the decrease in use of other herbicides.

GM soybean applications and the use of glyphosate is thought to support the use of conservation tillage. The US Conservation Technology Information Centre (CTIC) describes in its report an increase in conservation tillage in the USA. Increase has been significant since 1996 and mainly in crops for which herbicide-tolerant varieties are available. According to surveys with farmers, the main reason not to adopt conservation tillage is weed control. Conservation tillage decreases wind and water erosion and leads to better humidity and biodiversity in soil. In 2000, 12.6 million ha soybean were cultivated with reduced tillage, and 10.3 million ha with no-till cultivation. About 64% and 75%, respectively, were GM soybean planted areas.

Herbicide resistance of weeds is a well known phenomenon, dependent on the characteristics of the herbicide rather than the crop (long residual activity, single target site, specific mode of action, broad spectrum of activity and frequent applications with no rotation to other herbicides or other farming measures). Glyphosate has no residual activity and is degraded quickly in the soil, and as such is considered as a low-risk herbicide for resistance development. Until mid-2002 only three cases of glyphosate-resistant weeds had been identified, two of those stemming from before the introduction of GM crops. The third one, glyphosate-resistant marestail appeared within a 3-year glyphosate usage in GM-HT soybean in Delaware, USA. Weed resistance problems are not different from non-GM crops, giving importance to a proper weed management and careful use of herbicide.

Fernandez-Cornejo and McBride (2002) concluded that in 1997 and 1998 herbicide-tolerant soybean did not have a significant impact on overall farm net returns. Profitability was very much dependent on the specific weed pressure experienced by the farm. These results suggest that other factors might have a strong influence on adoption of GM soybean. Generally, the simplicity of the weed control programme using glyphosate and to be able to rely only on one herbicide for several weeds without harming the crop and influencing the crop rotation are the most cited reason to adopt GM soybeans (Carpenter et al. 2002). Savings for all soybean growers resulted from the fact that due to competitively priced glyphosate programmes other herbicide producers lowered their prices by up to 40%. This resulted in cost reductions of $216 million in 1999 compared to 1995, taking into account the technology fee for RR soybean (Carpenter and Gianessi 2001).

Marra et al. (2002) analysed several studies concerning the presented evidence for farm level impacts of GM crops. Regarding HT soybeans, most studies showed a slight reduction in yield. Only one profit estimation of $14/ha was available for North Carolina in 1997, which coincided with a yield increase for HT soybeans. The authors come to the conclusion that more research is necessary to be able to assess potential benefits, but that the broad adoption of HT soybeans clearly indicates profitability for most farm types as well as growing conditions in the

USA. Savings in herbicide costs and tillage costs will make up for lost revenue because of yield loss.

2.4.3.2
Herbicide-Tolerant Cotton

Cotton is a very important fibre plant, representing 33% of the total fibre market and 77% of the natural fibre market. The main producers are India, USA and China, with Greece being the ninth biggest producer and Spain being at place 22 in the list. GM cotton represented 20% of the global cotton acreage in 2002, 6.8 million ha out of 34 million ha (Fig. 2.2). Some 42% of GM cotton planted is herbicide-tolerant, 25% is insect-resistant Bt-cotton, 33% is cotton with combined traits (herbicide tolerance and insecticide resistance).

The USA accounts for about 20% of the world's annual cotton production, about 6.2 million ha were cultivated with cotton in 2001. In 2000, 72% of the national cotton area were planted with GM cotton (39% Bt-cotton, 61% HT cotton and 28% cotton with combined traits) (Fig. 2.3). There are two different GM herbicide-tolerant cotton varieties on the market in the USA: a cotton tolerant to the post-emergence herbicide bromoxynil (BXN cotton) and Roundup Ready tolerant cotton (RR cotton). Since its introduction in 1995 the area planted with BXN cotton increased to 8% of national cotton area in 1999 and decreased to 4% in 2001 (0.21 million ha). The decline is said to be connected to the boll weevil eradication programmes running in Tennessee and Arkansas, the two main cultivation areas for BXN cotton, and to increased demand for Bt cotton. Cotton varieties with both traits are not available. RR cotton was introduced in 1997 and increased to 60% of national cotton area in 2001 (3.7 million ha).

Weeds can reduce cotton yields considerably if not treated. Estimations indicate that without weed control, yields would decrease by 77% (Gianessi et al. 2002). More than 90% of the cotton area is treated with herbicides, and due to measures taken yield losses because of weeds are only about 7%. The typical cotton area (calculations including 15 of 17 cotton-growing US states, with Kansas and New Mexico missing with 52,600 ha) was treated in 1995 three times with three different active ingredients. On average 2.5 kg/ha active ingredient was applied with costs of $55.1/ha (14,690 t/year; $302 million/year).

Estimations for overall weed control costs additionally include herbicide application costs ($34.8/ha, $190 million/year), tillage ($42.4/ha, $232 million/year) and costs for hand weeding ($13.2/ha, $72 million/year). Overall costs are thus $797 million or $145.8/ha. Use of HT cotton in general does not result in better results of weed control compared to traditional treatment. The difference is in increased flexibility for weed management, including less costs and increased convenience.

Using different compiled data sources, and basing the comparison on base year 1997 for most of the 17 states that grow cotton, and the year 2000, the following differences were calculated (referring to 6.2 million ha planted cotton area):

– Herbicide costs: The average dosage of herbicide active ingredient used on cotton area decreased to 2.1 kg/ha (0.4 kg/ha or 2,800 t less) and costs per ha decreased to $47/ha ($47 million less were spent on herbicides).

Table 2.5. Summary of cost changes for herbicide-tolerant cotton concerning weed control

Herbicide costs million $/y	Application costs million $/y	Adoption cost million $/y	Tillage million $/y	Hand weeding million $/y	Total million $/y ($/ha)
–47	–58	+80	–53	–54	–132 (35)

- Tillage costs: Tillage has been reduced in nearly all cotton growing states with cost reduction of about $53 million.
- Herbicide applications have been unchanged or reduced. Two applications per ha are assumed, which leads to cost reductions of about $58 million.
- Hand weeding activities have been reduced with cost reductions of about $54 million.

Additional costs are based on the adoption costs for GM cotton, which vary from region to region. Adoption costs have been calculated according to planted area and specific regional technology fee: In 15 states RR cotton was grown in 2000 on 3.38 million ha with adoption costs of about $74 million. BXN cotton was grown on 386,000 ha with adoption costs of $6 million (see Table 2.5).

Carpenter and Gianessi (2001) stated a decline in herbicide use since introduction of GM cotton. However, they attribute the decline to a significant extent as due to the introduction and use of a 1996 introduced new post-emergence herbicide (Pyrithiobac, Staple), which is widely used at low average rates (0.06 kg/ha). Comparing 1994 to 1999, 1,735 t less herbicides were used with 1.3 million less applications.

2.4.3.3
Bt-Cotton

Bt-cotton was planted on 14% of US cotton area in 1996 (0.7 million ha) and increased to 34% by 2001 (2.1 million ha) (James 2002a). Most of Bt-cotton planted is also herbicide-tolerant.

The main insect pests for cotton in the USA are those that effect the squares and bolls: the bollworm, tobacco budworm, pink bollworm, boll weevil and lygus bugs. Without effective control the cotton bollworm and the tobacco budworm can cause average yield losses of about 67%. Pink bollworm might reduce yield by 50–80%. About 0.28 kg/ha active ingredient of insecticides are used against bollworm/budworm. Pink bollworm is combated by reducing the possibilities for overwintering through harvesting early and shredding and ploughing cotton stalks. Despite control measures yield losses can range between 4.5% and 11% (1995–2001, www.msstate.edu). Before the introduction of Bt-cotton, insecticides were used on 75% of the cotton area with 2.4 applications/year (Carpenter and Gianessi 2001).

Bt-cotton is protected against tobacco budworm, bollworm and pink bollworm. Research indicated that Bt cotton provides 70–90% control of cotton bollworm (bloom and prebloom), 95% control of tobacco budworm and 99% control of pink bollworm.

Cotton generally has very high production costs, in the USA it is only second to rice in per hectare costs of $1313 compared to $965/ha maize, or $657/ha for soybean. Approximately $148/ha are spent on chemicals (excluding fertilisers), compared to $71/ha for maize (USDA, ERS, 2002). About 25% of all pesticides used world-wide are applied to cotton (Carpenter et al. 2002), in the USA cotton is the second most heavily treated crop with about 37,000 t of pesticides applied each year.

An analysis of Bt-cotton performance in the different US states based on comparison between adopters and non-adopters in the same year (i.e. same conditions, infestations and insecticide programmes available) indicated that Bt-cotton led to an increase in production of 84,000 t of lint with a value of $115 million (national production in 2000 was 3.7 million t with a total cotton crop value of $4.8 billion). This reflects yield increases of 2% up to 23%. Costs increased by $12 million (technology fee), resulting in a net gain of $103 million or $50/ha. Estimations indicate that insecticide reductions directly connected to Bt-cotton use were about 900 t in the year 2000 (see Table 2.5).

Based on USDA data from 1995 to 1999, a significant reduction in pesticide use in six cotton-growing states could be shown (Carpenter and Gianessi 2001): 1,250 t less insecticides were used in 1999, accounting for 14% of all insecticides used in these states. Also the number of insecticide applications had been reduced by 15 million or 22%. On average a 9% higher yield could be obtained with a net gain of $20.8/ha.

Marra et al. (2002) in their analysis of several studies and surveys, which cover the time period of 1995 to 1998, came to the conclusion that data showed an average yield for Bt-cotton higher than for conventional cotton in 9 out of 11 cotton-growing US states. A reduction between 1.3 and 3 pesticide sprays per season led to reduction of average pesticide costs. The mean profit increase, taking into account the technology fee, was in the range of $50/ha to $247/ha. This is in line with the results presented by Gianessi et al. (2002), who calculated a benefit of $50/ha.

2.4.3.3.1
Development of Bt-Resistant Pests

A major concern, especially of organic farmers, is that cultivation of Bt-crops will lead to pest resistance against Bt toxins and thus make one of the few means to fight pests available to organic farmers ineffective. To avoid development of pest resistance, farmers need to grow a certain share of fields with conventional crops, either treated with insecticides or not. A recent study from Carriere et al. (2003) found that Bt-cotton enables a long-term control of pink bollworm. In 15 cotton-growing regions of Arizona the pink bollworm density was studied 5 years before and 5 years after introduction of Bt-cotton from 1992 to 2001. Results indicate that pink bollworm population densities were reduced significantly in areas of abundant Bt-cotton cultivation. The decrease was independent of demographic effects of weather and variations among regions. According to the authors, such long-term suppression has not been observed with insecticide sprays. The effect of long-term regional pest suppression might positively affect the refuge planting because of less pest pressure and less losses, thus reducing the risk of resistance development.

2.4.3.4
Bt-Maize

Maize is cultivated on 140 million ha world-wide. About 5.5% or 7.7 million ha were covered with Bt-maize in 2002, thus representing 13% of global GM crop area. Herbicide-tolerant maize is grown on 2.5 million ha and maize with combined traits (Bt and herbicide tolerance traits) is grown on 2.2 million ha.

In the USA, maize is the most important crop grown, with 32.4 million ha cultivation area representing 25% of all crops grown in the USA. About 43% of world production comes from the USA. Bt-maize was grown on about 6 million ha in 2001, thus representing 21% of the national maize area.

The main pests for maize in the USA are the European corn borer (ECB) and the south western corn borer (SWCB). The infestation differs across the USA with low, medium and high infestation zones, and also differs from year to year. According to annual surveys from USDA (United States Department of Agriculture) from 1942 to 1974, losses due to ECB varied between 838,000 t to over 7.62 million t/year (national maize for grain production in 2000 was 250 million t).

Not all farmers use insecticides against corn borers. Infestation levels generally vary unpredictably from year to year. Insecticides are only effective shortly after the hatching of the ECB larvae before they bore into the stalks and cannot be reached anymore by spraying. However, egg laying can occur during a time interval of 3 weeks and the insecticides usually are only effective for a short time period after spraying. To get the right time for applications, management time for crop walking and scouting has to be invested. It is estimated that less than 5% of the US Corn Belt maize area is treated with insecticides (Carpenter and Gianessi 2001).

Bt-maize was introduced to US agriculture in 1996. ECB as well as SWCB are susceptible to the Bt toxin produced by the plants. To calculate possible impacts of Bt-maize on yield and insecticide use, it was compared to production management using or not using insecticides. The analysis included 36 states with 99% of the US maize area. The results are presented in Table 2.6.

Based on data from Minnesota from 1983 annual losses of corn to uncontrolled ECB were estimated at more than $1 billion/year. Eight different active insecticide ingredients are recommended for fighting the corn borers. The costs of a single insecticide application are estimated at $34.5/ha.

For years with low infestation Bt-maize is compared to conventional maize production without insecticide application. For years with a high infestation Bt-maize is compared to insecticide application. Additionally the impact of Bt-maize adoption compared to a "typical year" has been calculated, based on the distribution of low and high infestation years over a 10-year period in the different US states and calculating the characteristics of an average year. Regarding the production volume increase it is assumed that insecticide application results in 80% loss reduction, whereas Bt-maize is assumed to prevent losses by 100%. In the calculation only the additional 20% yield increase due to Bt-maize compared to insecticide use are included as a benefit (Table 2.6).

Table 2.6. Impacts of Bt-maize cultivation considering low and high infestation years

Infestation level	Production volume increase million t/y	Production value increase million $/y	Production costs million $/y	Total net value million $/y ($/ha)
Low	+1.85	+145.3	−97 (Bt adoption costs)	48.3 (8)
High	+1.3	+102	−97 (Bt adoption costs) instead of −203.5 insecticides costs −106.5	208.5 (34.5)
Typical	+1.6	+126.5	1.1	125.4 (21)

Area 6.043 million ha.
Price $78.74/t.
Bt-maize technology fee $16.06/ha.
Insecticide application $34.5/ha, 0.43 kg/ha active ingredient.

During a low infestation year Bt-maize is assumed to increase yield by 1.85 million t with a value of $145 million. Taking into account adoption costs for Bt-maize this still leads to savings of $8/ha. Considering a year of high infestation, yield could increase by 1.3 million t with a value of $102 million. Additionally $203 million could be saved on pesticides (2,500 t less) resulting in net savings of $34/ha.

The estimations for an average year out of a 10-year cycle are calculated considering distribution of low and high infestation years throughout this period. Yield increases about 1.6 million t(about 3% yield increase) with a value of $126 million. Adoption costs are calculated against cost savings because of less pesticide use in high infestation years. Overall a cost reduction of $21/ha can be expected. Concerning insecticide application, Bt-maize might lead in a typical year to about 1,200 t less of insecticides. Marra et al. (2002) calculated on the basis of several studies covering the years 1997 and 1998 in the Corn Belt, an "unambiguous" yield increase for Bt-maize between 4% and 11% per hectare.

Carpenter and Gianessi (2001) analysed Bt-maize performance from 1996 to 1999 and compared it to conventional maize. The impacts of Bt-maize concerned mainly yield increases, e.g. in 1999 207 kg/ha or 1.7 million t, the equivalent of 200,000 ha harvest. Still, in spite of the costs of the technology fee, farmers earned about $4/ha less. 1998 and 1999 were years with a historically low pest pressure and it is estimated that in 10 out of 13 years, farmers would have gained from using Bt-maize. Regarding insecticides, the use of the five insecticides mainly used for ECB and SWCG, decreased between 1995 and 1999 by 6%. This reduction could have several, Bt-maize independent reasons, so that a maximum of 1.5% reduction is assumed to be due to Bt planting. Benbrook (2001a) concluded that insecticide applications targeted to ECB rose from 4% to 5% of area treated from 1995 to 2000.

Another benefit of Bt-maize is the reduced content of fumonisins due to less Fusarium infections. Infections with fungi are so-called opportunistic infections, taking advantage of lesions in the plant due to insect feeding. Mold fungi often contaminate stored food secreting bioactive secondary metabolites (mycotoxins), e.g. aflatoxins (*Aspergillus*) and fumonisins (*Fusarium*). Aflatoxins are liver toxins but are also considered as being carcinogenic. Fumonisins have been connected to esophagal cancer in humans and several animal diseases such as liver toxicity in poultry, hepato-carcinogenicity in rats, or congestive heart failure in primates.

Investigations have been carried out in Spain and France with Bt-maize Mon810 from Monsanto in comparison to a near-isogenic traditional hybrid (Bakan et al 2002), showing reductions in mycotoxin content for the Bt-maize. The fungal mass on Bt-maize was 4–18 times lower than on the isogenic maize; fumonisin concentrations ranged from 0.05 to 0.3 ppm for Bt-maize and 0.4 to 9 ppm for conventional maize.

2.4.3.5
Virus-Resistant Papaya and Squash

Papaya production is located in Hawaii on 648 ha with an annual production of 24,000 t ($17 million). The papaya ringspot virus (PRSV) is the main papaya disease world-wide. It reduced the production on Hawaii by 50%. Virus-resistant GM papaya became available with approval of federal regulatory agencies in 1998. Data for 2000 indicate that at that time 53% of the bearing area of Hawaii was GM papaya. From 1999 to 2000 the production increased by 33%, which is attributed to planting of GM papaya. It is estimated that GM papaya will soon be planted on more than 90% of the papaya area. The basis for the calculation of benefits is the assumption that planting of GM papaya will prevent the loss of 24,000 t harvest with a value of $17 million. This translates to $26,240/ha.

Squash is grown in the USA on 27,125 ha. Some 27% of the production is located in two states (Georgia and Florida) for which data on GM squash production is available (11,650 ha, 150,000 t production). For squash four mosaic viruses are important, with yield losses of about 20–80%, depending on when the crop is infected during development. According to data from the last few years an average yield loss of 20% can be expected. Viruses are transmitted by aphids so either insecticides are applied or oil application is used to prevent the aphids from feeding on the plant. Both solutions are considered to provide incomplete protection (estimated cost per application $25/ha).

Since 1998 two virus-resistant GM varieties, being resistant against three of the four viruses, have been available and planted on about 2024 ha, mostly in Georgia and Florida (17%). Based on the assumption of a 10% yield increase compared to oil/insecticide treatment (equal to 1.3 t/ha squash), the production on 2024 ha increases by about 2,600 t with a value of $2 million. Impacts on insecticide use have not been assumed as the insecticides used also control other pests apart from aphids. Adoption costs are estimated at $185/ha which results in a net benefit of about $1.6 million or $790/ha (Table 2.7).

Figures 2.4 and 2.5 show the calculated benefits on a farm level of the different GM crops analysed. Herbicide savings are biggest with HT soybean, ranging

Table 2.7. Impact of biotech crops planted in the US

Crop	Area included in calculations ha (% of national crop area)	Production/year			Total net value million $ ($/ha)	Pesticide use t
		Volume t	Value million $	Costs million $		
Soybean (HT)	20,256 (69)	0	0	−1,311 herb, +300 technology fee	1,011 (50)	−13,000
Cotton (HT)	3,767 (70)	0	0	−212, +80 technology fee	−132 (35)	−2,800
Cotton (Bt)	2,083 (34)	+84	+115	−12	103 (50)	−900
Maize (Bt) (typical year)	6,045 (21)	+1,600	+126.5	−1.1	125.4 (21)	−1,200
Squash (VR)	2	+2.6	2	0	1.6 (790)	0
Papaya (VR)	0.65	+24	17	0	17 (26,240)	0

Source: Gianessi et al. (2002).

at about 13,000 t/year or 0.65 kg/ha active ingredient. HT cotton application presumably results in herbicide reductions of about 2,800 t or 0.4 kg/ha. Bt-maize application results in 0.2 kg/ha less insecticide use (about 1,200 t less), assuming insecticide use in high infestation years. For Bt-cotton about twice the amount of insecticides could be saved, about 0.43 kg/ha active ingredient or 900 t.

Regarding profit increase, the best results are estimated for virus-resistant squash and papaya, $790 and $26,240/ha, respectively. These results stem from

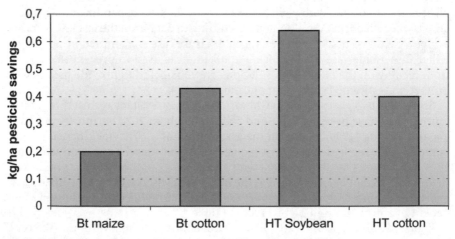

Fig. 2.4. Pesticide savings per GM crop (source: Gianessi et al. 2002)

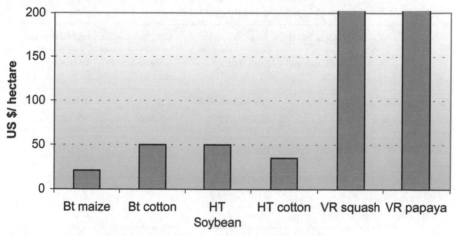

Fig. 2.5. Calculated savings in $/ha per GM crop (source: Gianessi et al. 2002)

the considerable yield increase and lack of any competitive conventional treatment. For the other crops analysed, the benefits range between $21/ha and $50/ha. For the herbicide-tolerant crops, benefits are based on savings on weed management, including fewer applications, less tillage etc. The benefits for Bt-crops result mainly from yield gains.

2.4.4
European Perspective

The only member state where GM crops (i.e. Bt-maize) are currently being grown on a commercial basis is Spain. Bt-maize is planted on about 20,000–25,000 ha, which represent about 4% to 5% of the grain maize cultivation area in Spain. About 90% of maize production is irrigated, with higher yields compared to dry land production. G. Brookes analysed the farm level impact of growing Bt-maize for Spain in 2002 (Brookes 2002). Up to now only one single Bt-maize variety is available, which is planted for feed purposes.

Bt-maize is used in areas with high to medium corn borer pest pressure. Generally 25% of maize planted in Spain is grown in high-pressure areas, about 40% in areas with medium pressure. Some 6–20% of Spanish maize is treated with insecticides to combat ECB, mainly by using chloropyrifos added to irrigation water (additional costs 18–24 €/ha) or aerial spraying (36–42 €/ha).

Yield losses are about 10–40% (average 15%) without using insecticides and about 5–20% with the use of insecticides (average 10%). The analysis of costs and benefits of using Bt-maize in the Huesca region (northwestern Spain) gives the results presented in Table 2.8. The farms analysed were about 50 ha or below in size (average in Huesca region was 50 ha). The Sarinena region experiences high infestations with ECB and the Barbastro region low to medium infestations. The adoption of Bt-maize in the Sarinena region, where insecticides have been previously used to combat ECB, leads to average yield increases of about 10% (1 t/ha).

Table 2.8. Average costs and benefits of planting Bt maize in two Huesca regions

	High infestation area; Sarinena		Low infestation area; Barbastro	
	Conventional maize	Bt maize	Conventional maize	Bt maize
Seed costs (€/ha)	150	168.5 (+18.5)	150	168.5 (+18.5)
Crop protection costs (€/ha)	144–222[1]	105 (90–120[2]) (–42)	90–120[2]	90–120[2]
Labour cost/ convenience	Benefit	Benefit	Benefit	Benefit
Yield (t/ha)	10	11 (10.5–12)	13–14.8	13.15–15
Yield gain (t/ha)		+1.0 (+123)		+0.15 (+18.5)
Net balance (€/ha)		146.5		0

[1] Includes 90–120 €/ha herbicide treatment, 24–102 €/ha insecticide treatment.
[2] Includes only 90–120 €/ha herbicide treatment. Source: Brookes (2002).

Less costs for insecticides result in overall benefits of about 146 €/ha. Yield increases in the Barbastro region with low to medium infestations are 0.15 t/ha (about 1%), resulting in no additional gains apart from the benefits regarding labour costs and convenience, which have not been quantified in the study.

Considering future prospects for the whole of Spain, about 36% of the total maize production area in Spain could potentially be planted with Bt-maize (173,000 ha), according to pest pressure and availability of the Bt technology for all leading varieties. Based on this assumption an average yield increase of 5–7% was calculated, which would result in additional production of 88,000–124,000 t or 1.8–2.5%. The reduction of insecticides used would be in the range of 35,000–56,000 kg active ingredient or 26–35%, with a reduction of area sprayed of 27–45%. The reduction in insecticide use could be lower, if insecticides would still be used for other pests (Table 2.8).

May (2003) studied the potential economic consequences for farmers for growing herbicide-tolerant sugar beet in the UK. Assuming that HT sugar beet is grown on 100% of the conventional sugar beet planting area (about 150,000 ha) average national savings of about 220 €/ha per year (14.7% of average costs) or 32.6 million € per year have been calculated. About 50% of the cost reduction stems from less agrochemical use.

Phipps and Park (2002) estimated reduction in pesticide use in Europe based on published data for other regions. Assuming that 50% of maize, oilseed rape, sugar beet and cotton would be GM varieties (herbicide-tolerant and insect-resistant), 4.4 million kg of active ingredient could be used less per year in Europe. About 7.5 million ha would be sprayed less, resulting in savings in diesel of about 20.5 million L. The basis for the calculation is the comparison with standard pesticide programmes identified for the different crops and the programmes for GM varieties in selected countries (Table 2.9).

Table 2.9. Potential reduction in pesticide use in Europe for GM crops

Crop	Area of GM crop[1] million ha	Pesticide reduction				Reduction in spray application	
		kg/ha	t	a.i. kg/h	a.i. t	Sprays/ha	million ha
Maize	2.2	1.6	3,520	0.82	1,800	1	2.2
Oilseed rape	1.5	1.0	1,000	1.1	1,100	2	3.0
Sugar beet	1.0	6.9	6,900	1.25	1,250	2	2.0
Cotton	0.25	12.2	3,050	0.84	210	1	0.25
Total	4.95		14,470		4,360		7.45

[1] 50% of European crop area is assumed to be planted with GM crops. Source: Phipps and Parks (2002).

A recent study from Gianessi et al. (2003) focuses on potential impacts of GM crops on European agriculture. At the time of writing three case studies on Bt-maize, HT sugarbeet and fungi-resistant potatoes had been published. The case studies for maize and sugar beet are mainly based on results from field trials and the experience gained with Bt-maize planting in Spain (see above). Table 2.10 provides a summary of the results. Regarding Bt-maize, the main grain maize producing European countries were covered by the exercise: France, Italy, Spain and Germany. The total adoption area is assumed at 1.599 million ha, including 40% of France's maize production area, 50% of Italy's, 36% of Spain's and 25% of Germany's, reflecting the areas which are highly infested with ECB. It is estimated that 28% of those areas is normally treated with insecticides, corresponding to 52.6 t active ingredients and costs of 13 million €/year. About 4% of the area is assumed to be treated with biological methods using the parasitic wasp *Trichogramma* with costs of 2 million €. Calculated yield increases because of use of Bt-maize range from 10% (compared to insecticide treatment) to 15% (compared to no ECB treatment). Increased production costs of overall 14.4 million € are contrasted by increased production volume of 1.9 million t or 263.4 million € and 52.6 t less insecticide use. The net gain per hectare of 156 €/ha is in the same range as calculated for Spain alone (Brookes 2002) (Table 2.10).

Regarding HT (glyphosate-tolerant) sugar beet, it is assumed that the adoption rate is 100% in each of the eight countries included, which totals 1.63 million ha (representing 88% of EU's 15 sugar beet areas). Sugar beets are usually treated with four to five applications of different herbicides, which corresponds to about 5,300 t active ingredients and costs of 331 million €/year (includes application costs). HT sugar beet planting is estimated to result in savings on costs for herbicide treatment of 181 million €/year or 111 €/ha. On average 1.3 kg/ha or 2200 t active ingredient less would be applied. The net income increase of 390 million € includes a yield gain of 5% based on the assumption that herbicide damage to the crop would be reduced by using glyphosate. A technology fee of 38 €/ha was assumed. The net gain calculated for UK alone is consistent with the results presented by May (2003) (see above). Reductions of herbicide use of 2,208 t are in the same range as the data presented by Phipps and Parks (2002).

Table 2.10. Potential impact of biotech crops planted in selected countries in the EU

Crop	Area included in calculations ha (% of specific crop area in the countries)	Production/year			Total net value million € (€/ha)	Pesticide use t
		Volume t	Value million €	Costs million €		
Bt maize[1]	1,599 (41)	+1,900	263.4	14.432	+249 (156)	−52.6
HT sugar beet[2]	1,628 (100)	+ 5,050	209	−181	+390 (240)	−2,208

[1] Countries included in the analysis: France, Italy, Spain, Germany
[2] Countries included in the analysis: UK, France, Germany, Italy, Netherlands, Belgium, Spain, Denmark.

2.4.5
Conclusion

GM crops have been introduced in agriculture in 1996. Since then adoption of this new technology has been widespread but very crop-specific, resulting in very high adoption rates for herbicide-tolerant crops such as soybean and cotton and less high adoption for insect-resistant crops such as Bt-cotton and Bt-maize. Also herbicide-tolerant oilseed rape in Canada reached very high adoption rates with over 50% of the national oilseed rape area. Apparently, the possibility of using a broad-spectrum herbicide without harming the crop, with certain flexibility, provides a higher advantage than the rather selective effect of Bt technology.

The high adoption rates, still increasing since 1996 speak for themselves, indicating strong benefits at the farm level. However, these advantages have not yet been completely quantified. Available studies point out that a comprehensive data base is still missing. Moreover, the interpretation of existing data has been subject to different approaches, leading to different results (Heselmans 2001). Marra et al. (2002) describe the possibilities for bias in collecting data on GM crop planting and economic impacts. The survey methodology used plays a significant role. Field-level surveys compare randomly selected single plots of randomly selected adopters and non-adopters of the technology (e.g. surveys of USDA). Differences in yield can thus only be calculated and compared as an average value for different varieties grown under different conditions. The yield difference calculated will be influenced by the comparator: Non-adopters can be less educated farmers with smaller farms and generally lower yields or farms with higher yields and less pest pressure. The difference will be larger or smaller. The results will differ from within-farm comparisons.

In addition to impacts that are accessible to quantification there are also impacts, that play an important role, but are difficult to quantify. The "convenience factor" belongs to this group. Less and more flexible application of glyphosate or glufosinate herbicide results in more time available for the management of other crops on the farm. Or, in case of Bt-maize, with often unpredictable infestation levels, the insurance effect against economic losses of growing a protected crop might influence the planting decision positively. The effect of Bt-crops depends on the pest pressure. If infestation is low, the benefit of growing a Bt-crop will be low compared to conventional crops. If infestation is high, the benefit will also be high, because of yield gains and less use of insecticides. This effect can be seen in the case of Spain. In the low infestation area presented, the moderate yield increase of 1% results in a break-even concerning additional costs for the technology fee but with the advantage of not having to check for infestation levels. In the high infestation area yield increases to 10%, which is higher than the yield increase (3%) assumed by Gianessi et al. (2002).

Compared to operating costs forecasted for 2002 by USDA, benefits from growing GM crops as calculated by Gianessi et al. (2002) can be substantial. In the case of HT soybean, savings of $50/ha represent 25% of the operating costs of $203.2/ha. For Bt-maize and cotton and HT cotton savings represent 5% to 7%, of operating costs. Cotton with combined traits has not been subject of the analysis but could enhance savings significantly. For squash and papaya savings are very high, because of the assumed yield gain.

Cost reductions are calculated without assuming any additional cost for segregation and identity preservation, as this is being done for most GM crops in the USA. Developments in the EU, considering a labelling threshold of 0.9% and legislation concerning traceability of GM crops, indicate that the situation will be different. Considering co-existence of different farming systems at the farm level, especially organic farming and GM crops, crop management measures need to be applied to minimise adventitious mixing (Bock et al. 2002), the additional cost depending on the crop and on the farm structure. To ensure segregation and traceability throughout the food chain, it might be necessary to develop distinct handling and supply chains.

Estimations for Europe show potential pesticide reductions of 0.88 kg/ha active ingredient, assuming that 50% of the maize, oilseed rape, sugar beet and cotton area would be planted with herbicide-tolerant or insect-resistant GM crops. Another study for Bt-maize and HT sugar beet calculated reductions of about 0.033 kg/ha and 1.3 kg/ha, respectively (corresponding to an average reduction 0f 0.69 kg/ha) (Gianessi et al. 2003). Low reductions for Bt-maize are based on the assumption that only 28% of the area is treated with insecticides. Gianessi et al. (2002), considering a more than 50% share of adoption for the USA, state average pesticide savings of 0.55 kg/ha active ingredient. Additionally, glyphosate and glufosinate are considered less harmful than other herbicides they replace. Glyphosate, due to rapidly binding to the soil, prevents leaching into the ground water and is biodegradable by soil bacteria with a half-life of 47 days in the environment (other herbicides 60–90 days). It is 3.4–16.8 times less toxic to mammals, birds and fish than other herbicides it replaces (Fernandez-Cornejo and McBride 2002). Thus, even if less herbicide is saved, the replacement effect already has a positive effect on human health and the environment.

However, the development of pest and weed resistance is a problem that users of GM crops also have to take into account. The refuge strategy (planting part of the crop area of each farm with conventional varieties) is used to prevent the development of Bt-resistant pests. Glyphosate-resistant weeds already exist, not exclusively due to glyphosate application in GM crops, and farmers must take care via thoughtful weed management strategies to prevent glyphosate or glufosinate from becoming a useless herbicide. A recent review on published scientific literature concludes that currently available evidence shows that geneflow from GM crops to weedy relatives occurs, but with low frequency. So far no evidence for adverse environmental effects of commercialised GM crops is available (ICSU 2003).

GM crops certainly are not the only means by which agriculture can be rendered more environmentally friendly, but they might contribute to that aim. The ICSU study (2003) concludes that GM soybean, corn and cotton provide agricultural management options for weed, insect and disease control that are consistent with improved environmental stewardship. Furthermore, GM crops could provide solutions to production security, safety and environmental benefits. Currently available GM crops provide benefits through enhanced conservation of soils and water, increased beneficial insect populations and improved water and air quality.

Yield increases because of less damage by pests or weeds could increase productivity per hectare thus limiting use of land for agricultural purposes, also with a view to increasing world population. GM crops that are currently under development in the EU show an on-going trend to agronomic input traits with an increased variety of GM crops. GM crops with a clear consumer advantage (e.g. hypoallergenic crops) are still further away from market introduction (unless via imports), probably also due to a significant slow-down in research activities in the EU (Lheureux et al. 2003). There is a clear need to act cautiously, assessing potential risks and benefits of GM crops carefully and to follow the overall aim of an environmentally benign agriculture.

2.4.6
Information Sources

Bakan B, Melcion D, Richard-Molard D, Cahagnier B (2002) Fungal growth and fusarium myco-toxin content in isogenic traditional maize and genetically modified maize grown in France and Spain. J Agric Food Chem 50:728–731

Benbrook C (2001) Troubled times amid commercial success for Roundup Ready Soybeans. AgBioTech InfoNet Technical paper no 4

Benbrook C (2001a) Do GM crops mean less pesticide use? Pesticide Outlook, pp 204–207

Bock AK, Lheureux K, Libeau-Dulos M, Nilsagard H, Rodriguez-Cerezo E (2002) Scenarios for co-existence of genetically modified, conventional and organic crops in European agriculture. IPTS Technical Reports Series EUR 20394 EN

Brookes G (2002) The farm level impact of using Bt maize in Spain. Brookes West, Canterbury, UK

Carpenter JE (2001) Comparing Roundup Ready and conventional soybean yields 1999. National Centre for Food and Agricultural Policy, Washington, DC

Carpenter J, Gianessi L (2001) Agricultural biotechnology: updated benefits estimates. National Centre for Food and Agricultural Policy, Washington, DC

Carpenter J, Felsot A, Goode T, Hammig M, Onstad D, Sankula S (2002) Comparative environ-mental impacts of biotechnology-derived and traditional soybean, corn and cotton crops. Council for Agricultural Science and Technology, Washington, DC

Carriere Y, Eller-Kirk C, Sisterson M, Antilla L, Whitlow M, Dennehy T, Tabashnik B (2003) Long-term regional suppression of pink bollworm by *Bacillus thuringiensis* cotton. PNAS 100:1519

ENDS Environment Daily (19 Feb 2003) More EU countries notify GM products. Issue 1388: www.environmentdaily.com

European Council and Parliament (2001) Directive 2001/18 of the European Parliament and of the Council of 12 March 2001 on the deliberate release into the environment of genetically modified organisms and repealing Council Directive 90/220/EEC. Off J Europ Commun 106:1

Fawcett R ,Towery D (2002) Conservation and plant biotechnology. Conservation Technology Information Centre CTIC, W Lafayette, IN

Fernandez-Cornejo J, McBride W (2002) Adoption of bioengineered crops. USDA Agricultural Economic Report No. 810

Fulponi L (2000) Modern biotechnology and agricultural markets: a discussion of selected issues. OECD, Paris

Gianessi LP, Sankula S, Reigner N (2003) Plant biotechnology: potential impact for improving pest management in european agriculture: www.ncfap.org

Gianessi LP, Silvers CS, Sankula S, Carpenter JE (2002) Plant biotechnology: current and potential impact for improving pest management in US agriculture: an analysis of 40 case studies. National Center for Food and Agricultural Policy, Washington, DC

Heselmans M (2001) Jury out on environmental impact of GM soy. Nature Biotechnol 19:700

International Council for Science (ICSU) (2003) New genetics, food and agriculture: Scientific discoveries – societal dilemmas. ICSU, Paris

James C (2002) Preview global status of commercialised transgenic crops: 2002. ISAAA Briefs 27

James C (2002a) Global review of commercialised transgenic crops: 2001. ISAAA Briefs 26

Lheureux K, Libeau-Dulos M, Nilsgard H, Rodriguez-Cerezo E, Menrad K, Menrad M, Vorgrimler D (2003) Review of GMOs under research and development and in the pipeline in Europe. IPTS Technical Reports Series EUR 20680 EN

Marra MC, Pardey PG, Alston JM (2002) The payoffs to agricultural biotechnology: An assessment of the evidence. International Food Policy Research Institute, EPTD Discussion paper no 87

May MJ (2003) Economic consequences for UK farmers of growing GM herbicide-tolerant sugar beet. Ann Appl Biol 142:41–48

Phipps RH, Parks JR (2002) Environmental benefits of genetically modified crops; global and european perspectives on their ability to reduce pesticide use. J Animal Feed Sci 11:1–18

US Department of Agriculture (2002) Economic Research Service http://www. ers.usda.gov/ Data/CostsAndReturns/data/Forecast/ cop_forecast.xls

2.5
Environmental Issues and Problems Associated with Genetically Modified Plants

Gert E. de Vries, Jaroslava Ovesna

Rapid advancements in biotechnology, along with the release into the environment and commercialisation of genetically modified plants have raised public and scientific concern over biosafety issues. In the European Union public attitudes towards genetically modified (GM) plants reveal a strong emotion oriented bias. Applications in agriculture are often perceived with mistrust. Ecological concerns about the utilisation of transgenic crops include the notion that engineered genes may be introduced in feral populations or persist as hybrids with wild species, resulting in the creation of superweeds. In addition, it is questioned as to whether or not commercial-scale use of transgenic crops might lead to gene transfer and recombination, creating new pathogenic bacteria or viruses. The impact of GM plants on the ecology of insect populations is also considered. Many aspects are therefore taken into account and studied by biosafety risk

sments, conducted prior to the release of GM plants into the environment.
... urope, risk management is based on the precautionary principle. Although
there are no major negative environmental impacts reported so far that can be
attributed to the use of GM plants more studies may be needed to support such
a conclusion.

2.5.1
Introduction

A gene from a firefly has been placed in a potato plant, making it light up whenever it needs watering. Rice plants have been genetically transformed to produce
a precursor of vitamin A, possibly helping millions of malnourished children
from going blind. Crops are engineered so that they can grow in contaminated
soil. Plants can be modified to produce plastics or pharmaceuticals. These are just
a few of the claims touted to justify the genetic modification of crops for the
benefit of mankind. Proponents believe that this technique's unprecedented
power has the potential to end world hunger.

GMOs are organisms that contain genetic information (usually one or more
genes) that enrich their genomes in a way that does not occur in nature. The
transfer of a distinct gene from one organism to another has now become a
routine procedure for many plant species. Genes of a large variety of microbes,
plants or animals, previously outside the gene-pool of a certain crop species,
can now be introduced, brought to expression, and tested. The first generation of
GM plants, which appeared in fields in the early 1990s, carried traits such as
herbicide tolerance or pest resistance predominantly to decrease inputs into crop
production.

Ag-biotech companies and some farmers stress the advantages of GM plants
in increasing productivity and benefiting the environment, and many scientists
believe that the new molecular tools for plant breeding have great potential. Yet
critics say that scientists are tampering with nature and, despite the commercial
success of GM crops in the U.S, they point out that there may be serious risks
associated with the release of GM organisms into the environment. Risks, real or
perceived, must always be balanced against benefits. In making value judgments
about risks and benefits when using modern biotechnological methods, it is
important to distinguish between technology-inherent risks and technology-
transcending risks. The former concerns food safety and the behaviour of a
biotechnology-based product in the environment. The latter originate from the
political and social context in which the technology is used and how these uses
may benefit and/or harm the interests of different groups in society in the
context of other current technologies used.

In this section the environmental risks of GM plants will be discussed and
compared to conventional crops that are the product of breeding and biotechnological methods that are viewed as safe and "natural" (see also Sect. 1.2
and 2.1).

2.5.2
Breeding and Genetic Modification

Many of the agricultural crops and other plant varieties with superior qualities that are currently grown in the world are the result of a long history of plant breeding and selection of the most valuable genotypes, with careful analysis of their progeny. The classical introduction of new traits (such as disease resistance) by crossing into otherwise well-performing varieties or the selection of efficient combinations of parental genes to ensure better end-use quality, usually takes over ten years. Therefore, in addition to the practice of conventional breeding, breeders have employed an impressive range of sophisticated methods to improve the speed of selecting desired genotypes. The latest of these tools is direct gene transfer, a method with increased precision that is combined with detailed information about the expected and resulting product. However, opponents of this technique view conventional methods as "natural" and criticise the use of gene transfer and genetic modification that provides us with the means to overcome the species barrier. Is it reasonable that the latter technique is heavily debated?

2.5.2.1
The Species Barrier

The practice of traditional breeding in general is limited to the transfer of traits between the same or fairly closely related species. The vast majority of conventional plant varieties possess, through controlled pollination, combinations of up to a hundred allelic variants of the same genes. Traditional breeding techniques rely on such new combinations of genes, which are selected from the available gene-pool of compatible plants. In contrast, GM technology allows the introduction of individual genes, even from unrelated species, resulting in genotypes with novel qualities. Molecular analyses can further determine the exact nature of the changes in the genome that have been created. Despite the great prospects, gene transfer between unrelated species has raised reservations with respect to ethics, environment, food safety and effects on world agro-economy. While it may be argued that natural barriers between varieties are also crossed when plant cells are fused using conventional tissue culture techniques (as in potato, maize, sugar-beet and others), it is the scale and range of novel opportunities that may concern individuals.

2.5.2.2
Genome Stability

Plant genomes, which determine the genetic make-up of plants as we know them today, have evolved with and without the help of man through duplication, variation, mutation, reorganisation by transposable elements, by uptake of DNA from the environment and by mutual exchange of useful traits. Genomes are not quiet and stable repositories of unchanging genetic information. How stable is the plant genome and what is foreign DNA? Is there evidence that the plant genome already contains DNA from external sources? Yes, indeed. In addition to the nucleus, plant cell organelles, such as the chloroplast and mitochondrion, contain genetic material which is thought to originate from archaic bacterial types. Some of this genetic information has even migrated from these organelles and can now be found in the plant nucleus.

Is there proof of the natural exchange of genetic material between unrelated organisms in the (distant) past? It is very likely that there is such proof. Highly similar DNA sequences and biochemical pathways can be found in completely different species, across genera, and even across phylogenetic kingdoms. Viruses may have acted as intermediaries in such processes of horizontal gene transfer. The frequency of occurrence is unclear but it may be a significant source of genomic variation in bacteria and possibly a route for evolution in eukaryotes.

Inter-specific crosses, such as sexual hybridisation between two individuals representing different species, also occurs in nature. In general such progeny is not viable, sometimes it is sterile or overall fitness is low compared to the parents. In inter-species hybrids of both mammals and plants, part of the genome may be activated and grossly amplified. Viable hybrid progeny have therefore apparently coped with massive and random disruptions of genome functions.

Traditional breeding techniques, mutation procedures, cell fusion or GM are all techniques that alter the genetic composition of an organism, with the goal of developing a new variety that offers improved traits. Plant breeding across species barriers (so-called wide hybridisation), or cell fusion techniques may cause genome disruptions on a far greater scale than the relative straightforward method of integration of transgenes by GM. The new cereal species *Triticale* is an example of a successful wide hybridisation. It is an intra-specific cross between wheat (*Triticum*) and rye (*Secale*), which are separate species from different plant families. Although the precise composition of the genomes of such new plant varieties are not well known, and the siblings may be quite unstable, these are not subject to extensive regulation and are not put to the same food safety and environmental impact tests as GMOs currently are. All the same, GMOs are subject to criticism and mistrust.

2.5.3
Environmental Risks

Several objections have arisen against the introduction of GMOs into the environment: genes could escape into other varieties and wild relatives, they could spread into other species such as bacteria, or they could affect non-targeted organisms. Is there scientific evidence for such fears?

There is concern that GM crop plants compromise the ecological values of natural habitats, increase the fitness of weeds when geneflow occurs, or that GM plant materials could play a role in the transfer of genes between species that are not usually sexually compatible. GM plants that carry genes for antimicrobial proteins or Bt toxins could also have adverse effects on the fitness of living organisms such as soil microbial communities. How differently do GM crops interact with other organisms when compared to conventional agricultural plants? (see also Sects. 3.5.3 and 3.5.4).

2.5.3.1
Plant–Plant Interactions

2.5.3.1.1
Geneflow

The fact that genes could spread from agricultural crops to wild relatives had not drawn much attention until the first series of field tests were performed with GM plants. However, since transgenic crops carry traits that may not be part of the natural gene pool of the inter-crossing plants, they are the subject of special attention. Geneflow is the natural process of exchange of genes due to pollen dispersal and is responsible for the spread of trait variants among different populations of the same or related species. Two issues are at stake: (1) can we protect non-GM cultivars from pollination by GM pollen, and (2) what risks are involved when transgenes become established in wild populations? The latter point is of special concern when it comes to the protection of local landraces, the natural genebanks of our current crops. If the genetic make-up of these original wild relatives were to change due to pressure from cultivated crops (GM or non-GM), valuable sources of original traits might be lost.

Different crops show different rates of self pollination and out-crossing. Some crops will hybridise with wild relatives, while others will not. British research showed that canola pollen spread at least 5 km and may cover distances up to 200 km by wind or insects. In other species (such as corn or wheat) these distances are significantly shorter. The characterisation of geneflow for the main crops is therefore an important issue in risk assessments (see Table 2.11).

The conclusion must be that if it is not possible to prevent the spread of transgenic pollen, transgenes will certainly spread, just as agronomic traits have spread to wild populations. The question is: does it matter?

2.5.3.1.2
Fitness

Wild plants, or weeds, have high fitness because they possess many or all of the following characteristics: stress tolerance, enhanced ability to make use of available soil nutrients, broad pest and disease resistance, early germination, rapid

Table 2.11. Frequency of gene flow throughout crossing

Crop	Crop to crop	Crop to wild relatives
Oilseed rape	High	High
Sugar beet	Medium to high	Medium to high
Maize	Medium to high	Medium to high
Potatoes	Low	Low
Wheat	Low	Low
Barley	Low	Low
Fruits*	Medium to high	Medium to high

* Strawberries, apples, grapevine, plums, raspberries, blackberries, blackcurrant.

growth, successive flowerings, early seed-ripening, high seed viability, and so on. Such traits are usually of a quantitative character and a specific combination of several genes determines overall fitness. It may not therefore be an easy step to transfer such traits between crops and wild relatives, let alone more distant species. However, if the properties of a single gene clearly affect stress resistance, and thus fitness, special attention should be devoted to the risk assessment. While annual crops such as cereals do not seem to be a weedy threat, perennial plants, including genetically engineered trees and grasses, could prove to be hardier than their wild counterparts.

If genes from GM crops cross out and into wild relatives, as genes from agricultural crops certainly do, a legitimate question would be whether novel traits would enhance the fitness of the recipient populations and lead to ecological changes. GM traits of interest are involved in stress response, fertility, and so on.

Herbicide-tolerant crops accounted for 74% of globally planted GM crops in the year 2001. A ten year survey of transgenic herbicide-tolerant maize, sugar beet, oilseed rape and potato crops planted by UK scientists concluded that herbicide resistance genes did not persist well in wild species and that herbicide-resistant GM plants are no more likely to invade other habitats than their unmodified counterparts. The herbicide tolerance trait is not likely to enhance fitness since there is no apparent selective advantage for plants carrying the trait.

The next generations of transgenic crops may be equipped with genes that will be responsible for increased crop yield or the production of specialty substances and it is expected that plants with such genes would exhibit a lower fitness. Indeed, in order to optimise yields farmers do need to support growth and protect agricultural crops from pathogens and far better competing weeds. However, certain traits could very well have a positive influence on fitness and cause such an effect when transferred to wild relatives. For instance, genes for increased tolerance to environmental stresses, such as cold or drought, or genes involved in pathogen resistance deserve special consideration. If such genes "escape" from crop plants into wild populations, their progeny could have a definite competitive advantage, potentially leading to an undesired shift in the existing ecological balance.

Even persisting crop plants may cause problems in natural environments. Rapeseed volunteers, not necessarily GM, are known examples. Such plants may become a weed on agricultural fields or disturb the landscape (headlands, roadsides, railways). It is clear that both the crop species in question and the type of transgene have to be taken into consideration in risk assessment. What can be dismissed is the general argument that GM crops will behave unpredictably on a long-term basis.

2.5.3.2
Plant–Insect Interaction

Insect attacks can cause severe damage in crops. Farmers have protected crops from these pests by using chemical compounds that are toxic to the target insect. The use of many such compounds that were used in the past has been discontinued because of human health or environmental reasons. GM technology has now allowed the development of insect-resistant plant varieties. The so-called Bt-crops contain a gene from the soil bacterium *Bacillus thuringiensis*, coding for a protein that happens to be toxic to a certain group of moths and insects. Such Bt-crops now can resist insect attack without application of pesticides. The technology did not come out of the blue – cultured micro-organisms have been used many years as a spray for the same purpose. While GM technology may reduce the use of chemical insecticides, widespread use of GM Bt-crops potentially introduces new problems that need adequate attention: the possible evolution of insect resistance to Bt, resulting in the depreciation of this natural insecticide, and the possible effects on non-target insects (the Monarch butterfly is an often cited example) or soil ecosystems. Additional information can be found in Sect. 2.2.

Possible impacts on non-target insect populations are usually carried out in feeding experiments to assess the risks associated with GM crops. Theoretical population dynamics and modelling are integrated parts of such research activities.

2.5.3.3
Plant–Vertebrate Interaction

Two types of animals may come into contact with GM crops or their products: farm animals may be fed with GM feed and wild animals may choose to take a bite from GM crops in the field. Currently approved GM crops are non-toxic; products are tested and so far have not led to any notable problems when used as feed or food. Therefore, harm to the natural fauna could only come from specialty crops grown to generate plant-derived pharmaceuticals or non-food chemicals with possible toxic properties. Such crops possess obvious risks for wild life and precautions should be in place when brought into cultivation. Humans will also be exposed to products from GM crops through their diet. However, there is no evidence that transgenic DNA acts any differently than the great number of DNA fragments that enter the human digestive tract with each meal.

2.5.3.4
Plant–Microbe Interaction

Movement of genetic information between sexually unrelated organisms, called horizontal gene transfer, is an important aspect of environmental impact assessments prepared by regulatory agencies. The possibility of a transfer of novel plant genes to taxonomically unrelated life forms, such as micro-organisms with possible consequences for ecosystems or human health, poses intriguing questions. Although horizontal gene transfer between plants and organisms of other kingdoms, such as micro-organisms, has not been aggressively explored, many experts are sceptical about the risks involved.

Many GM plants harbour antibiotic-resistance genes, since these marker genes were used in the process of the construction of the transgenic plant. There is concern that GM crop plants could render pathogenic bacteria unresponsive to antibiotics after receiving transgenes from GM plant materials. However, it must be realised that antibiotic-resistance genes in GM plants originate from micro-organisms in the first place. The traits are so widely spread among soil microbes that transfer of these genes from transgenic plants to soil microbes would not create novel gene combinations. Nevertheless, the revised EU regulation on the deliberate release into the environment of GMOs (Directive 2001/18/EC) includes the gradual phasing out of antibiotic-resistance marker genes by the end of 2004 for GMOs to be placed on the market and by end of 2008 in the case of field trials.

The current scepticism among scientists about the possibility of horizontal gene transfer is also fed by the fact that there is no recorded natural incidence, in the long history of agricultural experience, of a plant gene that has been put to work in a micro-organism. However, if we would hypothesise such a possibility, GM plant tissue could be a potential source for transgenic DNA sequences when released into the soil and taken up by soil microbes. Alternatively, horizontal transfer of a plant transgene into micro-organisms could take place in the digestive tract of animals or humans. It has proven difficult to show such horizontal gene transfer in normal situations since genetic material is rapidly degraded. While in controlled laboratory conditions genetic material from plant cells may be taken up by micro-organisms, the frequency of occurrence is considered too low to play a significant role in nature.

Horizontal gene transfer from bacteria to plants (not a subject for risk assessment) does exist in nature, in fact it made the first generation of GM plants possible. Note that bacterial genes will be silent in plants unless placed under a proper plant-specific regulatory system. The same is true for plant genes in bacteria.

2.5.3.5
Plant–Virus Interaction

It has been shown that transgenic plants, when containing a small but specific DNA fragment of a plant virus, may become resistant to infection by the target virus. This elegant method of protection against devastating crop losses due to viral infection may significantly reduce the use of agrochemicals. It is not known, nor can it be predicted, how fast viruses might overcome this type of resistance and leave GM plants once again susceptible to virus attack. As GM plants carry DNA information from plant viruses it is hypothesised (and shown in the laboratory) that exchange of information between live plant viruses and transgenic DNA may take place. This could lead to genetically modified plant viruses – possibly triggering accelerated plant virus mini-evolution. The possibility of such events must be examined carefully; the assessment of risks involved and the benefit analyses must be based on adequate scientific data sets. More detailed information can be found in Sect. 2.3.

2.5.4
Environmental Benefits

While there may be environmental risks involved when employing GMPs, there are also positive aspects that must be taken into account. Traditional agricultural practices may cause significant erosion of topsoil and sometimes may use precious water and other resources. It is expected that novel GM crops will be able to absorb nutrients from the soil more efficiently and, therefore, thrive in poor conditions, reduce the need for tillage and irrigation thus conserving soil structure and groundwater quality, and saving water. If GM crops achieve higher yields the continuing turnover of natural lands into cultivated area for agriculture may be slowed down. In addition, by overcoming local agronomic adverse conditions, such as frost, drought and soils with high salinity, the trend of an increased exploitation of tropical forests and other valuable natural reserves for agriculture may be curbed.

Insect-resistant crops have also shown a remarkable benefit for increased safety of foods. The absence of insect damage in GM crops due the presence of Bt endotoxins resulted in reduced fungal infection that otherwise may lead to the presence of carcinogenic mycotoxins in food products. Furthermore, a decreased use of chemical pesticides on GM crops that are insect, virus or fungal-resistant would certainly be a significant environmental benefit that hardly can be ignored.

2.5.5
Containment Strategies

Many of the environmental concerns about releasing genetically modified plants are connected with the fact that genes might be taken up in wild relatives, resulting in unwanted ecological changes. A foolproof system that would prevent such an escape of transgenic traits might contribute to solving this matter. Different routes have been taken to avoid geneflow via GM pollen.

Physical geneflow barriers are a practical way to prevent plant to plant gene transfer. This can be achieved by the use of isolation zones between GM and non-GM areas or by the planting of non-crop barrier fields that will trap any GM pollen. Another approach is to make sure that no transgenic pollen will be produced. These strategies include the blocking of floral development, male sterility, seed sterility, apomixis (production of seeds without fertilisation) and plastid transformation.

Male sterile plants do not produce pollen, and thus do not spread transgenes. Unfortunately for many crops it is important to produce seeds (cereals, maize, rape) so this approach cannot be followed. Sterile seeds may be produced through the use of "genetic use restriction technology" (GURT). This system makes use of a gene switch to "turn on" or "turn off" plant traits by the application of an external chemical substance. While this technology can be used at all levels of plant development, the switch can, for instance, be used to produce fertile seeds in otherwise non-fertile GM host plants. In this example seed companies would gain full control over the production of fertile seeds, while farmers would be restricted to producing non-fertile crop seeds as a product only.

Plastid transformation is another promising approach. Since pollen do not usually transfer plastids, such as chloroplasts, a relatively effective containment strategy would be developed if transgenes were introduced into the genomes of chloroplasts instead of the plant nucleus. Other advantages of placing transgenes in the plastid genome instead of the plant cell nucleus are: high expression and thus production of large quantities of speciality proteins and, on a technical level, the absence of gene silencing or positional effects and the possibility of introducing multigene cassettes. It must be noted that it will not be possible to utilise this strategy for the majority of plant nuclear genes that are involved in the main metabolic pathways.

2.5.6
Coexistence of GMOs with Other Agricultural Practices

Coexistence refers to the ability of farmers to provide consumers a choice between conventional, organic and GM products that comply with the European labeling and purity standards. Coexistence is concerned with potential economic loss through admixture of GM and non-GM crops or seeds, with identifying workable management measures to minimise admixture and with the cost of these measures. Because of the above-mentioned objectives, there is a need to develop measures, such as best practices and strategies, for the coexistence of GMOs with conventional and organic farming practices. The European Commission has published guidelines for the development of such measures. As with the labeling and traceability regulations, the guidelines are to be used in cases where there is no risk to human health or the environment.

The principles of these guidelines are based on:

- Transparency
- Scientific evidence
- Cooperation with all concerned, especially those at the farm level

The European Commission guidelines emphasise that developed measures need to be consistent with requirements as they relate to labeling thresholds and purity standards for GM food, feed and seed. The measures should be developed at a national and regional level and be cost-effective and efficient. They should be evaluated on crop to crop and region to region basis as the likelihood of admixture varies according to a number of variables such as seed impurities, cross pollination and volunteers (seeds remaining in the soil after harvest and producing new plants in successive years) (see Directive 2001/18). Appropriate measures and best practices must therefore be developed for cultivation, harvest, transport, storage and processing of material originating from the three different agricultural practices. The best practices for cultivation may include minimum isolation distances, physical buffer zones such as hedges or choice of crop types, and temporal buffer zones such as different reproductive times of crops.

A hypothetical situation where 10% or 50% of farms in a given region grow GM crops, revealed serious problems with the concept of coexistence. These problems arise due to different guidelines for GMO and organic farming practices. The adventitious presence of GM crops within an organic farm will vary according to the crop in question. Thus transgenic rapeseed, because of the problem with volunteers and ease of cross-pollination, would pose a impossible risk to an organic farm that does not allow any adventitious presence of GMOs. The indications are that at present only GM potatoes, as they are self pollinating and have no problems with volunteers, may be considered to be grown in the proximity of organic farms. Even that would still require changes in farm practices to avoid unintended admix of GM to non-GM potatoes in storage or transport. The estimation is that, with a hypothetical 10% or 50% share of the market, adventitious presence of GM potatoes in conventional or organic potatoes could be kept under 0.1% with great difficulty. Although this level is well below the requirements for labeling and traceability of conventional crops and foods (currently at 0.5%), it exceeds the zero limit set by organic farming (Regulation 1804/1999). It is thus unlikely that GM and organic crops can be grown in the same area in Europe. The liability for any damage to the price of the produce would lie with GM farmers (see also Sects. 4.1 and 4.2).

2.5.7
EU Regulations on Environmental Risk Assessment

It is clear that many issues play a role in biosafety considerations when employing genetically modified plants. Generally speaking, an ecological risk assessment is needed when a decision must be made as to whether a GM plant with a given trait can be released into a particular environment, and if so, under what conditions. If such releases are approved, the monitoring of the behaviour of the GM plants and their interaction with the environment after their release is a rich field for future research in crop ecology.

Biosafety risk assessments in the USA, which were conducted on the thousands of experimental field trials, have focused on the characteristics of the GM organism, the performance of its novel traits, the intended use of the organism and the risks associated with the recipient environment. Familiarity has emerged as a key biosafety qualification. Although familiarity cannot be equated with safety, it has provided a basis for applying existing management practices to new products. The concept is based upon a case-by-case and step-by-step risk assessment and management of novel products.

A recent development in Europe is the introduction of stricter regulatory requirements, based on the precautionary principle, partly in response to negative public reactions to increased use of GM plants in agriculture and food. This approach is based on the proposition that not enough may be known about possible long-term adverse effects of GMOs. The EU Directive 2001/18/EC regulates experimental releases and market introductions of GMOs and was adopted in the year 2001. The full implementation of this directive into national laws of EU Member States was still in progress in 2004. The EU regulation requires complex information to be provided on the specific GMO before its release into the environment:

– Taxonomic status and biology (such as the mode of reproduction and pollination, ability to cross with related species, pathogenicity) of the host organism
– Sufficient knowledge about the safety to human health and the environment of the parental (where appropriate), and recipient organisms in the neighbourhood of the release

- Information on any interaction of particular relevance for risk assessment, involving the parental (where appropriate) and recipient organism and other organisms in the experimental release ecosystem
- Information demonstrating that any inserted genetic material is well characterised
- The GMO shall not present additional or increased risks to human health or the environment under the conditions of the experimental release.

2.5.8
Information Sources

Biosafety-database is a scientific bibliographic collection of studies on "Biosafety and risk assessment in biotechnology". The database is updated monthly by the ICGEB. http://www.icgeb.trieste.it/~bsafesrv/bsfdata1.htm

Cartagena Protocol on Biosafety. The Biosafety Clearing House provides information that is relevant to the operations of the Biosafety protocol. www.biodiv.org/biosafety

EBSA (European BioSafety Association) is committed to enhancing the knowledge and understanding of biological safety issues throughout Europe and the world. http://www.ebsa.be

Scenarios for co-existence of genetically modified, conventional and organic crops in European agriculture (2002). European Commission Report EUR 20394

The Belgian Biosafety Server is the web server is hosted by the federal Scientific Institute of Public Health and keeps track of regulatory issues and risk assessment data. The site offers a range of background reading materials. http://biosafety.ihe.be/

TWN (third world network) page. The Third World Network is an independent non-profit international network of organisations and individuals involved in issues relating to development, the Third World and North-South issues. Biotechnology and biosafety is one of important issues http://www.twnside.org.sg/bio.htm

United Nations Environment Programme International Register on Biosafety. This web site offers information from many sources on biosafety. It focuses on information useful in establishing a regulatory framework for the safe development, transfer, and application of biotechnology. http://www.chem.unep.ch/biodiv/

3 Regulation, Assessment and Monitoring of GMOs

3.1
From a Research Proposal to a Product

Paul Pechan

Researchers usually study specific and detailed problems that have drawn their curiosity and belong to their field of expertise. Research is financed from various sources and usually includes resources supplied by the government and/or industry. Public funding usually covers specific and predefined research programs that are designed to attract competitive proposals. Therefore scientists will usually be required to prepare research applications that will be evaluated by expert panels. Application writing has become an almost full-time job of laboratory heads. Usually, a number of groups contribute to a research proposal. It is not unusual to write over 200 pages for a single research proposal.

Having succeeded in attracting funds, the conduct of research is monitored by various agencies and it may be required that scientists follow certain guidelines that include safety and ethical considerations. Examples of strictly controlled circumstances are the handling of GM organisms, pathogenic organisms, radioactive materials or the use of laboratory animals. As the size of the proposals grow, so does the need to make sure that the consortium of research groups agree on some basic cooperation principles that will be followed throughout the project duration. Such consortium agreements may also take a lot of time to prepare and can be highly complex, especially if industrial partners are involved.

Since public funding is paid for by tax money, it has been understood that the outcome of the research is shared with all others in the scientific community and utilised to the benefit of the public community. However, continuous reductions in public funding, the increased involvement of industrial parties and the calls for increased spending in applied sciences have changes these views. Scientists and public institutions are more inclined now than in the past to seek intellectual property protection for important scientific discoveries. Often, the proceeds of such activities will be re-channelled to the research groups where the invention took place.

New research results may be published in scientific journals. Before publication takes place, the results and any interpretations are scrutinised by fellow scientists. Through this process of review it will be determined whether the results are of sufficient quality and novelty to justify publication. The idea behind the necessity of publication is twofold. First, other researchers must have the opportunity to check the results and continue to develop the published findings. Second, publications are one way of measuring the productivity and quality of a given group or individual. With the publication of scientific results the knowledge enters the public domain, and anyone, including industrial parties, may try to put the knowledge to work and develop new products that can be marketed. However, if researchers protect their invention by filing for a patent, third parties would need to pay for licensing rights to further work and to market the research findings.

Industry support is usually directly invested in research groups that have gained a reputation or have shown a special expertise in a specific area of science. Industry funded research typically concerns problems that are linked to promising applications. The contractors usually stipulate that although the patent rights may be held by the research group, any rights to the exploitation of new discoveries will be owned and developed by the sponsoring industrial partner. If the product makes it to the market place, usually a very difficult and uncertain task, the research group/organisation would receive a certain percentage from net sales.

In industrial settings, research and development is usually divided into various phases. In the first phase research is carried out to verify and prove the concept of the discovery. This discovery may originate from internal research activities or from collaborations with public institutes. During a second phase, the research findings are refined and adapted to the needs of the company. Throughout this process an eye is kept on competitors to see whether similar research is being done and what progress is being made. During the development phase, scientific results will lead to real products. Questions regarding costs of production and marketing will finally determine whether the product will be developed or shelved.

All products must be tested and their safety assessed before introduction onto the market place. In the case of GM crops, extensive glasshouse and field trials as well as health safety tests are carried out before crops can be approved for environmental release. Safety considerations are very important, as are the approval procedures and consistent legislation. If the product is exported, it needs approval by the appropriate agencies around the world. This process is complicated by the fact that different countries have different standards for product approval. At the end it may be that some products that industry would like to place onto the market may have delays in approval or be rejected. A lot of invested money and time may thus be wasted. The risks associated with developing a new product is borne by industry. Thus one successful product may need to pay for many other products that have failed to make it onto the market.

Products based on GM plants are much more closely scrutinised than products of traditional plant breeding. They have an especially tough time making it onto the market. In contrast to plant breeding products, GM plants need to be shown to pose no hazard to public health and the environment. In Europe, such new varieties are given only a limited time permit for general release. All food products made, containing or composed of GMOs need to be labelled and traced as such. This adds costs and complicates marketing of GM products.

3.2
Genetically Modified Products: Intellectual Property Issues

Gert E. de Vries

For many years the western world has recognised life forms as patentable. This started in 1873 when Louis Pasteur obtained a US patent for pure yeast culture, as a composition of matter. Then in 1926 a patent was obtained for a bacterial fermentation process which produced butyl alcohol and acetone from maize, and in 1935 a patent was granted for a process which manufactured yeast. Patents must be novel, involve an inventive step and serve an industrial applicability. The Plant Breeders Variety Rights (PBR) Act provides protection for new plants and does not require either the method of producing the plant or the plant itself to be novel, inventive or otherwise fulfil the requirements of patentability, but provides encouragement in the form of protection for the results of conventional biological plant breeding. The rapid advancement of biotechnology and new trends in the field of biotechnology allows for a duel protection system of both patents and PBR relating to the same invention.

3.2.1
Introduction: History and Controversies

The goal of all plant breeders is to produce better performing plants. A legislative environment that provides protection for the results of breeding new plant varieties is an essential guarantee for innovation. An alternative method of protection is the development of technology that would prevent farmers from saving seeds and thus avoiding royalty payments. Protection is necessary since it would not make sense for a commercial operation to invest time and effort in producing better crops, if bankruptcy looms in the near future.

The so-called "plant variety protection" legal system is the usual choice to make a profit from the creation of new crop plants with enhanced properties. Patents are considered to be a better route when protecting individual traits or methodologies that allow the creation of novel plants with enhanced properties. The patent holder, in theory, is able to effectively block competitors from using plant varieties with special features and may exclusively exploit the discovery. Another method of protection relies on gene technology preventing the expression of traits unless activated by proprietary chemicals, to be supplied by the patent holder.

Driven by the return on investment, the public benefits from a wide range of innovations that might not have otherwise been available. In addition, the information disclosed in a patent is placed in the public domain and this has the benefit of publicising the current direction and extent of biotechnological research, which in turn stimulates further research and the development of more innovative products. Controversies on the effects of intellectual property protection arose when molecular genetics was added to the toolbox of plant breeders and the impact of genetic engineering on plant breeding became clear. Opposition exists against patent protection and the increased control by multi-national seed and agrochemical corporations resulting in loss of the farmers' independence and a possible limitation of agronomic gene pools. General objec-tions involve the patentability of DNA sequences and life forms. This section covers these various issues, the controversies and the consequences of the protection of inventions in plant sciences.

3.2.2
Intellectual Property: Protection Systems and Return on Investment

Intellectual property (IP) is a general term for a combination of acquired knowledge, innovation, experience and hard work. Protection of IP, by obtaining exclusive legal rights for trade or application, may be gained by any of the following legal provisions: copyright, trade mark, utility models and, of importance here, patents or plant variety protection.

3.2.2.1
Plant Variety Protection System (PVP): Scope of Protection

Specific protection for new plant varieties is created in accordance with the laws applying in the various member states as well as by Regulation no. 2100/94 (EC), the EC protection of new plant varieties (Community plant variety right) which is based on the revised Act of the Union for the Protection of New Varieties of Plants (UPOV), a Geneva-based intellectual property convention. The plant variety protection system exists alongside patent law.

A new plant variety is defined by the expression of individual characteristics meeting criteria such as: new, distinct, uniform, stable. These criteria are assessed during two years of on-site trials conducted by the granting office in conjunction with plant breeding institutions. Plant variety rights can be obtained for a national territory or at the EU community level.

Plant variety rights do not give protection equivalent to those afforded by patents. Plant variety rights contain both a breeder's/research privilege (a right to freely use a protected variety for the purpose of producing a new variety) and a farmer's privilege (the right to use farm-saved seed from one year to the next; under the Community Plant Variety Rights Regulation, however, this is subject to specified fees and regulation and for some crops is even restricted. The breeder's privilege is also restricted since a plant variety right includes "essentially derived varieties", in order to prevent the creation of cosmetically different varieties as a way of evading royalty payments.

3.2.2.2
Patents: Description, Exclusions and Fees

Europe has adopted Directive 98/44/EC on the Legal Protection of Biotechnological Inventions, which, in accordance with the WTO-TRIPS, states that member states must provide patent protection for all inventions, irrespective of the field of technology. An important exception is the exclusion of plant varieties. However, the Directive has so far been implemented by only a few countries and is under intensive discussion.

Patents are granted by regional patent offices such as the European Patent Office (EPO) or by national patent offices. As soon as a European patent, in essence a bundle of national patents, has been granted by the EPO it becomes a national patent in any or all of the 25 member states that were explicitly designated in the patent application. A patent request comprises of the application, which includes a description of the invention, one or more claims that define the scope of protection provided by the patent, any diagrams necessary to clarify the description or the claims and a brief abstract summarising the invention. The detailed description of an invention is published by patent offices in the form of patent specifications in patent documents. The scientific or technical know-how that made the invention possible is now made public, which contributes to the state of the art.

The basic criteria of patentability in all areas of technology include novelty (not state of the art), an inventive step (not obvious to experts) and industrial application (purpose and reproducibility). Discoveries therefore cannot be patented because the act of invention requires some process. Essentially biological processes, except those in the category of microbiology, also fall under this patenting exclusion. Therefore, conventional breeding procedures, which are wholly within the natural limits of crossing and selection and do not require human action, should be treated as essentially biological processes. On the other hand the patenting exclusion does not apply if physical or chemical procedures are involved, such as pruning or inventions concerning genetic manipulation of plants, even though the invention includes biological processes. Inventions which would be contrary to morality or that may offend the public are also excluded.

The mere sequencing of a genome belongs to the area of discovery and for that reason alone cannot take advantage of patent protection. It is different if a DNA sequence is released from its natural surroundings by means of a technical procedure and is made available for the first time to a commercial application. Here there is a step taken from knowing to being able to. However, when patenting the material that has been isolated, it is essential that its special function or useful characteristics can be defined and that the industrial application of a sequence or partial sequence is disclosed in the patent claim.

The US Patent and Trademark Office has released new guidelines that should resolve the controversial and ambiguous practice of patenting genes (⟨www.access.gpo.gov/su-docs/aces/fr-cont.html⟩). The description of mere genetic sequences or pieces of genes is not sufficient for a patent application since it was commented that a person whose body includes a patented gene could be guilty of patent infringement. However, when the inventor also discloses how to use the purified gene isolated from its natural state, the application will satisfy the "utility" requirement.

In recent years the European Patent Office (EPO) has made a number of fundamental decisions for the interpretation of the terms "plant varieties" and "biological processes for the production of plants". The EPO has awarded a patent to Aventis for plants that are genetically modified to resist its glufosinate herbicide and rejected an appeal filed by environmental group Greenpeace since the gene was modified and could not be found in nature in this form. For certain GM plants it may therefore depend on the inventor's choice to seek protection through PVP or by applying for a patent.

The total fee for the application of a patent will be made up, depending on the type of invention, of a number of necessary services: the initial claim submission, filing and search fees, examination fees, grant fees, translation costs and the agent's fees. Indeed one of the main factors that makes the present European system so expensive is the need to translate the patent specification into all the official languages of the member states. A patent holder remains responsible for payment of renewal fees, which, as a general rule, increase over time. This means that only the most commercially viable patents will be maintained for the full period of 20 years. The European Commission proposed, on the 5th of July 2000, the creation of a Community patent to give inventors the option of obtaining a single patent legally valid throughout the European Union with a principal aim of reducing the cost of patenting an invention in Europe. However, in May 2004 the EU ministers failed to agree on a EU-wide Community Patent, since there was no consensus on the number of languages to be used in the patent descriptions.

3.2.2.3
Licences: Types

Patents give the patentee the right to prevent third parties from making, using, or commercialising the invention during a maximum period of 20 years. If the owner of an invention does not have the necessary means to undertake the manufacture of the product resulting from his invention, the patent may be sold or handed over to achieve commercial exploitation of his property right. Exploitation may also take the form of licensing and can take the form of a single payment, or, be granted against royalties, which are linked to specified criteria such as sales volumes of the licensed product. In contrast to the sale of a patent, the licensor remains the owner of the patent and will remain responsible for its maintenance fees. A good understanding of how to manage and license intellectual property in the public sector is required to decide when inventions should be sold, licensed exclusively or licensed non-exclusively.

3.2.2.4
Infringement: Case History

Neither computer chips patented by Intel nor Round Up Ready soybean seed patented by Monsanto are public goods. If third parties use a patented invention without authorisation, the owner has the right to prohibit that use and to claim for damages. However, enforcement of these rights is likely to be much more difficult in the case of biological technology such as improved seed, because, contrary to computer chips, seeds multiply and the farmer may decide to use his own seed in future planting without paying the original owner. Although farmers may enter into contracts with seed companies agreeing not to use their own seed, such contracts are difficult to enforce.

A test court case unfolded in Canada where Monsanto used private investigators, who took samples from fields and harvests, to find that canola farmer Percy Schmeiser was using herbicide-resistant seed without permission. Schmeiser fought back with a $10 million lawsuit of his own in which he accused Monsanto of libel, trespass and contamination of his fields with Roundup Ready, by pollen flow from neighbours.

If a patent holder in Europe wishes to bring an action for infringement of, or challenge the validity of, a European Patent, it may be necessary to bring actions in a number of member states. This requirement means that litigation involving European Patents is expensive. The EU Commission has proposed amending the European Court of Justice such that disputes concerning Community Patents are dealt with by a single new centralised Community tribunal.

3.2.3
The Patent System and Society

The harmonisation of the different protection mechanisms of intellectual property in the world is a long and meticulous process since economic, social and cultural values are at stake.

3.2.3.1
Discrepancies in Legislation in Europe and the US

European Union licensing of GM products and the granting of patents has stalled in recent years because of perceived health and environmental concerns. After the failure of a first attempt, the European Directive (98/44/EC) on the legal protection of biotechnological inventions was adopted in July 1998 to harmonise national patent laws of the member states of the European Union.

A number of discrepancies exist between European and US patent laws and these are not only limited to dissimilar procedures:

1. Under the US patent system a plant variety, as the result of traditional plant breeding, is patentable. However, it must be noted that such a patent would prevent others from making, selling or using such a particular invention in the USA.
2. In Europe neither DNA nor raw genome information can be patented because they are discoveries and not inventions describing an application. In the USA the question as to how explicit DNA sequence functions must be described is not yet clarified. At present the possibility of patenting certain DNA sequences exists in the USA.
3. In Europe patent offices must make an assessment of whether an invention is contrary to morality before a patent can be granted.
4. In Europe certain farmer's rights have been retained in a similar way to those given in the Plant Variety Rights Directive. For specified species of fodder plants, cereals, potatoes, oil and fiber plants, small farmers have a right (except when the variety in question is a hybrid or synthetic) to use harvested products for propagating purposes in their own fields.
5. A six month grace period allows patent applicants in the USA to publish their results prior to the submission of the patent claim. Under the European Patent Convention almost any public disclosure of an invention prior to the priority date of a European patent application destroys the novelty of a patent application for that invention.

3.2.3.2
Company Strategies and the Patenting of Life

In plant sciences the patent system attracts special criticism, since life, which is considered public property, is an essential element in the invention. Also it is often held that all biological inventions, which deal with human, vegetable, or animal genes, involve materials which already occur in nature and can therefore under no circumstances be invented, but only discovered. However it is clear that when novel, primary gene sequences are modified and put to work in a different setting, the requirements of novelty, the inventive step and industrial application have been fulfilled.

Another recurring point of criticism are the methods used by companies to secure return on investment. Since patent rights are difficult to enforce, especially in the field of agriculture, technological methods have been developed to discourage the farmer from using saved seeds for the next year's plantings. For many years hybrid seeds have played such a role: while such F1 generations produce higher yields, further generations (saved seed) have lost this property. Gene technology allows a wider range of protection systems. The Technology Protection System (TPS, sometimes called "suicide genes" or "Terminator"), causes seeds to produce sterile plants unless pretreated by a proprietary chemical. The so-called trait-specific T-GURTs (genetic use restriction technologies) have the potential to affect variety of other traits; the goal is to turn a plant's genetic traits "on" or "off" with the application of an external chemical. While critics point at the increased control by multinational seed and agrochemical corporations, there are also clear benefits in spite of moral and social dilemmas: germination control technology provides a way to prevent the spread of genes introduced into improved crops, patented seeds could also be distributed in countries without a creditable system of patent protection, and farmers could be given the choice as to whether to activate specific traits.

3.2.3.3
Third World Countries: Private and Public Research

With the world population forecast to reach 7.5 billion in 20 years, there is a strong argument that GM crops have the potential to increase yields and produce the required amounts of foods locally. At the same time it is disturbing that patents with broad claims and restrictive contracts will put the price of seeds out of the range of Third-World nations.

In an effort to break this barrier scientists from seven international science academies launched a campaign supporting GM food and called for research to increase yields of tropical crops. The academies' report urges companies and research institutions to share knowledge and called for a ban on broad patents.

In another effort to get around the financial problems of Third-World nations, collaborative research projects are being conducted between European and their African/Asian laboratories to create disease-resistant transgenic plants. Crops such as cassava or banana do not have great priority in research programmes of agro-companies, but still are major crops in these countries. Transgenic varieties are therefore more likely to be improved and distributed by co-operating public research institutes. Failure by the public sector to expand investment in agricultural research will result in lost opportunities for increased economic growth and reduced poverty and food insecurity. So far, unfortunately, this is the current prognosis for most developing countries.

Agro-companies take initiatives as well. Monsanto launched a web site (http://www.rice-research.org) opening its rice genome sequence database to researchers around the world, facilitating the use of its technologies and data for the common good. The inventors (Potrykus/Beyer) of pro-vitamin-A enriched (golden) rice made their technology freely available to developing countries. Creation and production of the GM golden rice varieties would involve as many as 32 companies and institutions holding 70 patents that cover the necessary technologies. However, both Monsanto and Syngenta have pledged their support and will provide royalty-free licences for all of their technologies that can help further the development of golden rice.

3.2.4
Information Sources

The internet is an excellent resource for detailed information on IP, PVP, patents and how to apply for them, infringement cases and consequences for trade and development. This section therefore concludes with a list of essential resources.

3.2.4.1
Organisations and Agreements

CPVO   Community Plant Variety Office implements and applies the system for the protection of plant variety rights. The system allows intellectual property rights, valid throughout the Community, to be granted for plant varieties. http://www.cpvo.fr/en/default.html

EPO   In 1973 all EU Member States signed the European Patent Convention which established the European Patent Office (EPO) and a single procedure for granting patents. The organisation is independent of the EU and has non-EU members. http://www.epo.org or http://www.european-patent-office.org/index.en.php

EU   CommissionEuropean Directive (98/44/EC) for the protection of bio-technological inventions, search site. http://europa.eu.int/eur-lex/en/index.html

GATT   General Agreement on Tariffs and Trade, the outcome of the 1986–94 Uruguay Round negotiations, is the WTO's principal rule-book for trade in goods. The agreement created new rules for dealing with trade in services, relevant aspects of intellectual property, dispute settlement and trade policy reviews. http://gatt.org

PCT   Patent Co-operation Treaty was agreed upon in 1970 and has been ratified by 100 countries, including all those of the developed world. http://www.mewburn.co.uk/patsintf.htm

TRIPS   Agreement on Trade Related Intellectual Property Rights of the WTO covers the many aspects patent protection, notably that member states must provide patent protection for all inventions irrespective of the field of technology. http://www.southcentre.org/publications/trips/toc.htm

UPOV   Act of the Union for the Protection of New Varieties of Plants, revised in 1991, provides a definition of Plant Variety Protection for its member states. http://www.upov.int

USPTO   US Patent and Trademark Office. http://www.uspto.gov

WIPO   World Intellectual Property Organization (WIPO) is an inter-governmental organisation, headquartered in Geneva and responsible for the promotion of the protection of intellectual property throughout the world. http://www. wipo.int/index.html.en

WTO   Global harmonisation of trade and intellectual property provision is sought through the World Trade Organisation. http://www.wto.org

3.2.4.2
Other Internet Links

ETC Group is dedicated to the conservation and sustainable advancement of cultural and ecological diversity and human rights. ETC is concerned about the loss of genetic diversity – especially in agriculture – and about the impact of intellectual property rights on agriculture and world food security. http://www.etcgroup.org/main.asp

IBM Intellectual Property Network Web site allows patent searches on line. http://www. ibm.com/ibm/licensing

Information about the Plant Variety Protection Act in the USA. http://www.ams.usda.gov/science/PVPO/PVPindex.htm

Intellectual Property Mall Page, links to a large collection of intellectual property resources. http://www.ipmall.fplc.edu/

IPR helpdesk as a source and guide to patent information. http://www.cordis.lu/ipr-helpdesk/en/home.html

NAL/USDA Biotechnology Information Resource, a fairly extensive bibliography of articles and monographs related to biotechnology patents http://www.nal.usda.gov/bic/Biotech_Patents/

Patent primer. http://www2.ari.net/home/foley/page1.html

UK Patent Office home page. http://www.patent.gov.uk/

University of Wales, Aberystwyth, links to patent offices, patent search and download sites and other patent related sites. http://users.aber.ac.uk/dgw/patent.htm

UPSTO patent full-text database manual search, 31 field search capabilities such as: patent claim(s), description/specification, classification, parent case information. http://164.195.100.11/netahtml/search-adv.htm

Yahoo patent links. http://dir.yahoo.com/Government/Law/Intellectual_Property/Patents/

3.3
GMO Traceability and Labelling Regulations

Paul Pechan

GM plants and GM food are primarily regulated in the EU primarily on the basis of Directive 2001/18/EC and Regulations 1829/2003 and 1830/2003. The legislations stipulate a case-by-case risk assessment and a step-by-step development and testing of GMOs. Traceability of GMOs and labelling of GM feed are a new development, as is the labelling of all products as GMOs even if they no longer contain GM DNA and proteins. There is a 10 year limit on the market authorisation of GMO products. The European Food Risk Authority will play a central advisory role in the assessment and authorisation process. In USA, the legislative approach is product-based, not process-based as in the EU.

3.3.1
GMO Regulatory Framework

Since the beginning of the 1990s, a framework of European regulations have been in existence concerning GMOs. This framework was revised recently. Its main components are the horizontal Directive 2001/18/EC on the deliberate release into the environment of genetically modified organisms, which covers both the experimental release of GMOs and placing them on the market; Regulation (EC) 1829/2003 on genetically modified food and feed; and Regulation (EC) 1830/2003 concerning the traceability and labelling of genetically modified organisms, as well as food and feed products produced from GMOs. These regulations surplant Directive 90/220/EEC and Regulation 258/97/EC.

Another important feature of the regulatory framework is the European Food Safety Authority (EFSA), established in 2002, and – as well as addressing other food and feed safety related issues – this is the body responsible for independent scientific advice about and risk assessment of GMOs to be released into the environment and to be placed on the market as food or feed. This includes environmental risks and risks for human or animal health and safety. EFSA aim is to act proactively in order to identify emerging risks in the food chain. However, risk management, including appropriate decisions, are the responsibility of the European Commission in collaboration with the European Parliament and the Council.

The use of GMOs in food and feed requires authorisation according to Directive 2001/18/EC and Regulation (EC) 1829/2003. This includes a risk assessment and the provision of data showing the safety of the food or feed for human and animal health. With the revision of the regulatory framework, the "one door-one key" principle was introduced. Therefore, there will be only one risk assessment and one authorisation for the deliberate release of GMOs into the environment, such as the cultivation of GM maize, and its use as food or feed; for example GM sweet maize for human consumption or GM maize gluten for feed. These applications will be assessed by the EFSA. Information on authorised products will be available via a public database. Authorisations are valid for a period of ten years, and they are renewable.

The latest legislations are primarily aimed at increasing the public confidence in the regulatory and enforcement agencies in Europe as well as improving the reliability, efficiency and transparency of the decision-making process. In addition, traceability is viewed as part of the cautions approach to health and environmental safety as advocated by the European Commission's Precautionary Principle (see Sect. 4.2). In all cases, the products or processes allowed on the market are deemed to be safe for human or environmental use.

This short section concentrates on the traceability and labelling regulation. For overview of other EU legislations on GM foods, please read König et al. 2004 or see section 3.5.

3.3.1.1
GMO Traceability

EC regulation 1830/2003 introduces requirements for ensuring the traceability of GM products throughout the entire production and distribution chain from farm to store. This will enable post-market monitoring of potential adverse environmental and human health effects and, if necessary, the withdrawal of a product from the market. Traceability of GMOs is ensured by placing obligations on business operators to transmit and retain information at each stage when introducing a product onto the market. The traceability regulation requires that information about the GMO is passed along the production chain, from the farm to the marketplace.

3.3.1.2
GMO Labelling

All materials, such as food and feed, originating, containing or consisting of GMOs will need to be labelled as containing or having been produced from GMOs (EC Regulation 1830/2003). The unintended or technically unavoidable presence of approved GM material in non-GM products will be tolerated up to a threshold limit of 0.9%; above this they will also need to be labelled. This threshold level is calculated based on the proportion of genetically modified DNA or protein in the final product. This calculation may get complicated if a product is composed of two or more GM components. In such a case, the threshold level is calculated for each component separately. Products that contain GMOs that have not yet been authorised for market consumption are not allowed to be marketed in the EU. An exception is made for products that contain not more than 0.5% of a GMO that has received a positive scientific risk assessment. These products can be placed on the market, provided the presence of this GMO is adventitious. In principle, labelling is intended to inform consumers and enable them to choose whether to buy GM food or not. Products such as meat or eggs from animals that were fed with GM feed do not have to be labelled. Also, enzymes that are used in the production process and that are produced from GM microorganisms will not have to be labelled.

3.3.2
GMO Regulatory Framework in the USA

The EU approach to GMOs differs significantly from the USA, where the regulatory framework is based on the assumption that GM products do not pose risks different to those from similar, conventional products. For that reason, new specific regulations were deemed unnecessary, and the overall approach is product-based, not process-based like in the EU, where the process of genetic engineering triggers the application of specific legislation (see Table 3.1). Oversight authority should only be exercised if the risk posed by the new product is

Table 3.1. Differences between EU and US GMO product approval and labelling procedures

	EU	USA
Public opinions broadly consulted (as for example in labelling requirements)	Yes	Partly
Testing for environmental effects completed before market introduction, completed before marketing approval	Yes	No (can be done after introduction)
10 year limit on market approval	Yes	No
Broad environmental sustainability issues considered during risk assessment	Yes	Partly
Product traceability requirements	Yes	No
Case-by-case approach	Yes	No (trait based)
Post-marketing monitoring of environmental effect	Yes	No
Use of substantial equivalence	No	Yes

considered to be unreasonable. Regulatory oversight over GM products is shared by three agencies: The Environmental Protection Agency (EPA), the Food and Drug Administration (FDA), and the Animal and Plant Health Inspection Service (APHIS) of the US Department of Agriculture.

3.3.3
Information Sources

EC MEMO 02/160 and 03/186 provide questions and answers about the regulation of GMOs in the EU. http://europa.eu.int/rapid/start/cgi

EC Scientific Committees. The website provides access to information on activities and opinions of the eight Scientific Committees of the European Commission. http://europa.eu.int/comm/food/fs/sc/index_en.html

EUR-Lex provides free access to EU legislation in preparation and in force. See http://europa.eu.int/eur-lex/en/index.html. For GMO related issues see also: http://europa.eu.int/comm/food/fs/gmo/gmo_index_en.html

European Commission. The website provides access to the web pages of all Directorate Generals and as well according to subjects such as food safety. http://europa.eu.int/comm/index_en.htm

European Food and Safety Authority. Information on the Authority can be found at http://www.efsa.eu.int/index_en.html

European Parliament and Council (2001) Directive 2001/18/EC on the deliberate release into the environment of genetically modified organisms. Off J Europ Commun L106:1

European Parliament and Council (2003) Regulation (EC) No 1829/2003 of the European Parliament and of the Council of 22 September 2003 on genetically modified food and feed. Off J Europ Commun L268:1

European Parliament and Council (2003) Regulation (EC) No 1830/2003 of the European Parliament and of the Council of 22 September 2003 concerning the traceability and labelling of genetically modified organisms and the traceability of food and feed products produced from genetically modified organisms and amending Directive 2001/18/EC. Off J Europ Commun L268:24

König et al (2004) Genetically modified crops in the EU: food safety assessment, regulation, and public concerns. Office for Official Publications of the European Communities

National Academies. Homepage of the four US academies and institutes with an extensive list
   of searchable publications: http://www4. nationalacademies.org/news
US Department of Agriculture, Animal and Plant Health Inspection Service APHIS, Agricultural Biotechnology. http://www.aphis.usda.gov/ppq/biotech/
US Department of Health and Human Services, Food and Drug Administration, FDA. www.
   fda.gov
US Environmental Protection Agency. www.epa.gov

3.4
Plant Biotechnology and its Regulation in Central and Eastern Europe

Paul Pechan, Ervin Balazs

In the last two decades, many Central and Eastern European (CEE) countries have
emphasised the development of competitive plant biotechnology research programs with the expectation of enhancing agricultural production. The political
changes in the nineties were accompanied by serious economical problems. Although both the funding and the management of R&D was negatively affected,
many research groups have become internationally recognised for their work. In
the late 1990s several research institutions became centres of excellence, as judged
by the European Union. However, the results of plant biotechnology have not been
readily transferred to agriculture. Many newly formed agriculture-oriented companies and state-owned enterprises, that could have used plant biotechnology,
were mismanaged and went out of business when state subsidies were stopped.
Family-owned farms were not capable or interested in using these technologies.
Moreover, the management and application of the new biotechnologies was ham-
pered by the lack of appropriate biosafety regulations. The CEE countries that have
recently joined the EU are now changing their biosafety regulations as part of the
alignment process with the EU rules and regulations. The progress in creating and
enacting these regulations varies from country to country. Some differences also
exist in terms of how the regulations are supervised and how they are enforced.
    This section presents an overview of the recent plant biotechnology research
and regulations in the three new EU member states most active in the genetic
engineering of crops: Czech Republic, Hungary and Poland.

3.4.1
Historical Perspective

From the end of the Second World War up until late 1980s, CEE countries behind
the iron curtain were dominated by the former Soviet Union. The economy and
research were among the areas where the Soviet influence was most felt. Basic
research was concentrated in academic research institutes whereas tasks in
universities were reduced primarily to teaching. Applied research was performed
at special institutes run by governmental ministries.
    Although science was relatively well supported in several CEE countries, the
money was not used efficiently. Research funds were derived from state budgets.
Institutes, rather than projects, were supported. Political involvement was an

important criteria in personal advancements and it was expected that senior scientists would be members of the communist party. Travel and research exchange to the western world was restricted and under strong political state control with some exceptions especially in Poland and Hungary where scientists were able to travel abroad more easily than from other CEE countries.

Internationally, classical plant breeding was well recognised and respected. The breeding institutes, which usually included experimental stations, could be listed as success stories.

However, only with the opening up to the West, has plant biotechnology become firmly established in CEE countries. This has meant more research and with it the need to establish appropriate regulations governing genetic modification of plants. The need for proper regulations was accelerated by two additional factors: firstly, the need for CEE countries to harmonise regulations with the EU and, secondly, international companies became interested in field testing GM plant varieties in Central Europe.

3.4.2
Current Plant Biotechnology Research Situation in CEE Countries

The political changes of the nineties opened the doors for effective collaborative research. Changes to the funding systems, with emphasis on peer review of individual projects, helped to strengthen good research groups. With the start of the European Commission (EC) sixth framework program, scientists from CEE countries are fully integrated into the EC granting system. However, it is still very difficult to maintain scientific capacity in the region and offer young scientists exiting job prospects in order to prevent them from seeking employment elsewhere.

In as far as plant biotechnology research is concerned, the Czech Republic, Hungary and Poland are the most advanced countries among the CEE countries. Transgenic plants have been produced at many of the research institutions and field trials carried out. Plant biotechnology in these countries covers both classical biotechnology such as fermentation technology, tissue culture, micropropagation, cell biology, marker-assisted breeding as well as using molecular techniques to create GM plants.

3.4.2.1
Czech Republic

The research focused on plant biotechnology has been principally localised at institutions of the Czech Academy of Sciences, and to some extent at Charles University and the University of Olomouc. Applied research is carried out at the Research Institute of Crop Production (RICP) (belonging to the Ministry of Agriculture) and Agritec Ltd., Šumperk (formely also Ministry of Agriculture).

The Institute of Experimental Botany is divided into the Institute of Experimental Botany in Prague and Olomouc and the Institute of Molecular Plant Biology (IMPB) in České Budějovice. IEB is involved in research leading to plants with possible practical uses: for example potatoes with reduced sweetening during storage, development of plant vaccines against papiloma adenovirus (causing cancer in woman) or barley and wheat with improved feeding qualities. RICP is involved in several types of applied projects dealing mainly with crop improvements (for example potato with higher regeneration ability and potatoes with increased resistance towards abiotic stresses) and field evaluation of transgenic cultivars and biosafety. Agritec has employed genetic modifications to develop transgenic cultivars (flax, pea).

National research is supported by several agencies in the Czech Republic. National Agency for Agriculture research is run by the Ministry of Agriculture. It supports mainly applied agricultural research. Grant Agency of Czech Republic funds basic research in different disciplines. At the moment, there is some support for research projects aiming to develop transgenic plants with potential added values. Currently, biosafety projects are preferred.

3.4.2.2
Hungary

Szeged Biological Research Center (BRC), the plant breeding institute at Martonvásár (both part of the Hungarian Academy of Sciences) and the Agricultural Biotechnology Center (ABC) (Ministry of Agriculture), are the primary institutions where plant biotechnology research is carried out.

Genetic manipulation of haploid cell cultures, including anther cultures, and the in vitro fertilisation of isolated reproductive cells is at the forefront of research effort. A particle bombardment device (the Genebooster) has enabled transformation of several kinds of plant cells and tissues including rice, wheat, barley, species of poplar and carnation. Cell and tissue culture studies with maize and wheat concentrate on plant transformation and regeneration to develop aluminium tolerance. Cross protection against viruses in potato and tobacco has been developed using the coat protein transgene approach.

Fundamental research is mainly supported by the National Science Foundation (OTKA). Introduced in 1986, it is a peer-reviewed competitive grant system. Applied research is sponsored by the Bureau for technological development and is also based on competitive grant application system initiated in 1990.

3.4.2.3
Poland

In 1987 Andrzej Legocki from the Institute of Bioorganic Chemistry of the Polish Academy of Sciences established a group of molecular biologists and scientists to work on in-vitro plant culture. The task of the group was to adapt available transformation methods and the molecular procedures for transgenic plant production. A number of plant biotechnology groups were subsequently established at various institutes around the country.

Transgenic potato plants have been regenerated that are resistant to leafroll luteovirus at the Institute of Biochemistry and Biophysics of the Polish Academy of Sciences while transgenic cereals (triticale and rye) with herbicide resistance,

and virus resistance have been produced at the Plant Breeding and Acclimatiza-
tion Institute in Radzikow. At the Bioorganic Chemistry Institute transgenic
lettuce has been obtained expressing the hepatitis B virus protein gene. An efficient method for transformation of *Gerbera hybrida* plants has been established
at the Institute of Pomology and Floriculture in Skierniewice.

The research is supported by several agencies in Poland, mainly by the State
Committee for Scientific Research. Applied agricultural research is partially
supported by the Agency for Restructuring of Agriculture run by the Ministry of
Agriculture and Rural Development. The State Committee for Scientific Research
coordinate basic research in different disciplines such as agriculture and molecular biology. At the moment, research projects aiming at the development of
transgenic plants are well subsidised.

3.4.3
Regulations Governing GM Crops in CEE Countries

Under the Convention of Biological Diversity, several CEE countries agreed on
the implementation of appropriate biosafety mechanisms and on the creation of
national biosafety regulatory frameworks. In general, the elaboration and initi-
ation of the national biosafety frameworks were driven and based on EU directives, especially 90/219 and 90/220. Relevant laws were adopted in the different
countries by the Parliament only recently. Consequently, their implementation,
including enforcement, is not well advanced. In addition, their laws need to be
updated to be in line with the EU regulations from 2003 (see Sects. 3.3 and 4.1).
As a general rule, advisory bodies in CEE countries examine each request related
to handling of GM crops and inform the competent authority. NGOs are involved in the advisory bodies and the public is informed in advance about field
releases. Detection and monitoring of GMOs is carried out primarily by special
reference laboratories that participate in EU organised ring trials.

Risk assessment and management of GMOs are still dependent on pre-existing
national infrastructures and safety systems.

3.4.3.1
Hungary: A Case Study of National Biosafety Regulation

Hungary has been one of the pioneers for the establishment of biotechnology
legislation in the CEE region. The first gene technology Act was adopted by the
Hungarian Parliament on 16 March 1998. This legislation was based on EC
Directives 90/219 and 90/220. On 1 January, 1999, the Hungarian Gene Technology
Act (Act No XXVII of 1998) came into force.

The objective of the Act is to regulate activities in the field of modern biotechnology in order "to preserve the balance in nature, to protect human health,
to support scientific and economic development and to enforce the provisions
of the Convention on Biological Diversity".

The Act applies to the contained use, release, commercialisation, import
and export of GM organisms. Modification of human genes and genome does not
fall under this Act, but is covered by the Act on Public Health.

The Government is currently reviewing this regulatory system on the basis of their experience and with a view to the new EU regulations. The current key elements of the Hungarian Biotechnology Law are listed below.

3.4.3.1.1
Authorizing System

A permit is obligatory to:

- Establish a biotechnology laboratory
- Modify a natural living organism
- Use GMOs in a contained system
- Release GMOs into the environment
- Commercialise GMOs on the market
- Export and import GMOs

3.4.3.1.2
Evaluation

A semi-independent biotechnology committee prepares the decisions and gives opinions. It consists of 17 members representing competent ministries and many others (for example the Hungarian Academy of Sciences, National Committee on Technological Development and non-governmental organisations).

3.4.3.1.3
Issue of Permit

This is the responsibility of biotechnology authorities under the control of the competent ministries according to the industries where application of GMOs are intended. The authorities give the permits, control the application, in certain cases restrict or ban the GMO-activity, revoke the permit, or impose a fine. The authorising responsibility is shared by three ministries: Ministry of Agriculture and Regional Development, Ministry of Health, Ministry of Economics, in accordance with the most common applications of biotechnology.

Biotechnological (including genetic engineering) activities are divided into three categories by the law. These are the following:

- Plant and animal breeding, food and feed production
- Human health care, medicine production
- Industrial use not included in the preceding two categories

The Ministry of Environmental Protection also has an important role as a consultant (with a veto-authority) in giving an expert-opinion on the activities to be permitted.

3.4.3.2
Czech Republic

The Czech Advisory Committee for transgenic plants was established by Czech scientists in the early 1990s. Until the law covering GMOs was passed in 2000, the

Ministry of Environment, which had the primary responsibility for GMOs, consulted with this committee about the applications and releases of transgenic plants.

In May 2000, the Czech Parliament adopted an Act 158/2000 "Act on the use of genetically modified organisms and products and amendment of some related Acts". The Act offers the basis for the implementation of EC Directive 90/219/EEC, as amended by Directive 98/81/EEC, EC Directive 90/220/EEC, including the amendments agreed in June 1999 by the EU Council of Ministers, and the Cartagena Protocol on Biosafety. The Act is being implemented by two or more decrees, prepared by the Ministry of the Environment.

The Act covers contained use, deliberate release into the environment, and placing on the market of GMOs and products containing or consisting of GMOs. The main responsibility for the Act is with the Ministry of Environment, which established a Committee for GMO and derived product handling. Two other ministries are involved in handling the contained use, release into environment and deliberate releases of GMOs. The law is under further revision for harmonisation with the latest EC regulations.

3.4.3.3
Poland

In Poland, based on the initiative of the Ministry of Agriculture and Food Industry, an interdisciplinary consultative group on GMO was established in 1996 for consulting legal regulatory issues and for assessing applications for deliberate releases. The principle of the Polish national biosafety system is build upon existing institutions.

The Act on GMO went through several readings and was adopted by the Parliament in May 2001. The Law was published and came into force in October 2001, and will be updated on the basis of the latest EU regulations. The Law is under the responsibility of Ministry of Environment in cooperation with the Ministries of Health and Social Welfare, Ministry of Agriculture the Scientific Research Committee and other relevant ministries. The established Committee for GMOs have members from the representatives of the ministries and a group of experts. This group acts as an advisory body for the decision makers. Their main responsibility is to prepare recommendations for risk assessment and evaluation of the applications.

The competent authority considers GMOs in three sectors of activities, namely contained use, deliberate releases and commercialisation. Contained use and deliberate releases, regulated in the Polish law, are in harmony with the two corresponding EU Directives Modification of human genes and the human genome does not fall under this Act.

There are several state agencies which are responsible for the control measures and inspections (plant protection inspection, market inspection, customs service, environmental protection inspections, and so on). Reference laboratories provide the technical support for biosafety related issues.

3.4.4
Field releases and Commercialisation of GMOs in CEE Countries

Until 2003, only experimental field trial releases have been authorised by the competent authorities of the CEE countries. Examples on field trial releases authorised in the CEE countries are:

Hungary   Herbicide and insect-tolerant corn, herbicide-resistant oilseed rape, virus-resistant potato and tobacco, improved protein in wheat
Poland   Herbicide-resistant sugar beet, virus-resistant potato, herbicide-resistant oilseed rape, virus-resistant plum,
Czech   RepublicVirus-resistant plum, potato with reduced sweetening, improved flax

3.4.5
Information Sources

Biosafety regulations in CEE countries. All important latest documents on Biosafety and
    Regulation and texts of National Laws are available at http://www.biosafety-CEE.org
European Federation of Biotechnology (1999) Biotechnology legislation in Central and Eastern
    Europe, Briefpaper 9

3.5
Genetically Modified Plants and Risk Analysis

Anne-Katrin Bock

Risk analysis includes risk assessment, risk management and risk communication. Risk assessment is the first and crucial part of the risk analysis process of GMOs. The principal approach on a case-by-case basis and proceeding step-by-step is generally accepted, but harmonisation of the different methods used on an international basis is needed. Risk assessment needs to comply with high scientific standards. Scientific uncertainty in assessing potential risks needs to be acknowledged and dealt with in an open and transparent way that also includes the public. More research is necessary to fill some of the knowledge gaps.

3.5.1
Introduction

Risk assessment has a long tradition in regulating human activities with the aim to minimise or avoid risk to human health and the environment. Examples can be found in the production of medical products, chemistry or nuclear power. According to European regulations, the safety of GMOs has to be assessed prior to releases into the environment and placing on the market. The approach is described in more detail in Directive 2001/18/EC on the deliberate release into the environment of GMOs, which was adopted in April 2001 and repealed Directive 90/220/EEC in October 2002. In the Annex II of this Directive the principles for the so-called environmental risk assessment, which also includes human health effects, are laid down. Concerning food, Regulation (EC) 258/97 on Novel Food and Novel Food Ingredients stipulates risk assessment for foods that have not been used for human consumption to a significant degree in the European Union before. Foods and feed containing or consisting of or derived from GMOs are covered by Regulation (EC) 1829/2003, requiring one single risk assessment, carried out by the newly founded European Food Safety Authority. The overall aim is to release only those GMOs that do not pose any risk to human health or the environment. Possible positive effects of GMOs are not subject to risk assessment. Please also refer to König et al. (2004) for a thorough review of food safety assessment and regulation.

3.5.2
What is Risk Assessment?

Risk assessment, the first part of risk analysis, is followed by risk management and risk communication (see also Sects. 2.5, 3.6, 4.2 and 4.5). Environmental risk assessment is defined by Directive 2001/18/EC as the evaluation of risks to human health and the environment, whether direct or indirect, immediate or delayed, which experimental deliberate release or deliberate release by placing GMOs on the market may pose.

Direct effects refer to primary effects, which are due to the GMO itself, e.g. allergenicity of the derived novel GM food. In contrast, indirect effects occur through a causal chain of events, e.g. interaction with other organisms or effects due to a change of agricultural management due to the use of GM crops. Immediate effects could be observed during the period of release of the GMO, e.g. the establishment of weedy GM plants outside the agriculturally used fields. They can be direct or indirect. Delayed effects would be observable at a later stage as a direct or indirect effect as such as long-term effects from changed consumption patterns due to GM food. Additionally, cumulative long-term effects on the environment and human health have to be assessed.

The objective of environmental risk assessment, according to European legislation, is to identify and evaluate potential adverse effects of a GMO and to elucidate if there is a need for risk management and suitable measures to be taken.

In the context of this section, the terms hazard and risk are defined as follows: A hazard is a potential harmful characteristic (here of a GMO), which is an intrinsic property of the organism investigated. Hazards can give rise to negative consequences. These consequences can have different orders of magnitude and different likelihood of actually coming true. Risk can be quantified by combining the likelihood of consequences of a specific hazard with their magnitude.

The principal approach to assess the safety of GMOs is largely accepted. First of all risk assessment should be science-based and carried out ensuring a very high scientific standard. For every GMO the risk assessment is done on a case-by-case basis and in a stepwise manner. This means that for example each GM plant is tested first in the laboratory then on a small scale in a field trial, followed by a large-scale field trial before authorisation for placing on the market can be requested. The following step can only be carried out if the preceding step has shown that the GMO does not pose any risk to human health or the environment.

In contrast, the interpretation and use of the results of the risk assessment differ within the European Union Member States and internationally, depending for example on the models used for comparisons. For example, Germany and the UK compare the use and the effects of GM crops to conventional agriculture, while Austria or Sweden take an organic-oriented input reduced agriculture as the scale.

3.5.3
How is Risk Assessment Carried Out?

The steps in environmental risk assessment are outlined in Table 3.2. Potential adverse effects on the environment and human health depend strictly on the specific characteristics of the GMO and thus to a certain extent on the inserted transgene(s) and the respective traits. Potential hazards associated with GM crops are listed in a general way in Table 3.3 and will partly be explained in the following sections, distinguishing between environmental hazards and hazards for human health.

3.5.3.1
Spreading of the GMO in the Environment

What is the degree of invasiveness of conventional crops, and can transgenic traits increase the potential of survival in non-cultivated surrounding areas or as volunteers on the same plot? Many GM crops developed today carry herbicide tolerance as a new trait, which is not expected to increase the fitness of the plants in the absence of the selecting factor, i.e. the respective herbicide. The situation might be different when new traits such as increased tolerance to dryness, salt or a reduced need for nutrients are developed (see also Sect. 2.5.3).

3.5.3.2
Vertical and Horizontal Gene Transfer

The transfer of transgenes from GM crops to other related crops or weeds (vertical gene transfer, out-crossing) is a very intensively studied and discussed issue.

Table 3.2. Steps in environmental risk assessment

1	Identification of characteristics that may cause adverse effects
2	Evaluation of the potential consequences of each adverse effect if it occurs
3	Evaluation of the likelihood of the occurrence of each identified potential adverse effect
4	Estimation of risk posed by each identified characteristic of the GMO
5	Application of management strategies for risks from the deliberate release or marketing of the GMO
6	Determination of the overall risk of the GMO

Table 3.3. Potential hazards associated with GM crops

Expression of toxic or allergenic compounds	Potential for production of substances that are toxic or allergenic to human beings or other species
Effects on biogeochemistry	Potential to negatively influence decomposition processes in the soil and thus causing changes in nitrogen and carbon recycling
Increased persistence on the environment and invasiveness	Potential to confer an ecological fitness advantage to the GM crop causing persistence and invasiveness (superweeds)
Transfer of genetic material	Potential to transfer the newly introduced genetic material to other crops or weeds via cross-pollination or to other organisms via horizontal gene transfer. Depending on the transferred trait such gene transfer might not present a hazard
Instability of genetic modification	Potential of reversing down-regulation of a naturally occurring hazardous trait
Unintended effects	Potential that genetic modification leads to unintended effects, e.g. influencing other genes of the organisms, which might lead to unexpected hazards.

The risk of gene transfer to related weed species depends very much on the GM plant itself. Maize and potato do not have any compatible indigenous related weeds in Europe that could receive transgenes via pollen flow. In contrast, oilseed rape is a cross-pollinating species for which several related species exist, so out-crossing cannot be ruled out. The extent of out-crossing depends on climatic conditions, agricultural practices, viability of pollen, and availability of out-crossing partners. The establishment of a trait in the wild population depends on the selective advantage the new trait might confer. The possibility of gene transfer within the same crop species depends on the specific crop. It can present a potential problem for agriculture, as in the case of organic agriculture where only very low levels of GM plants might be tolerated in the harvest.

The term horizontal gene transfer describes non-sexual gene transfer e.g. from plant to micro-organisms. Micro-organisms, especially bacteria have the ability to take up DNA from other organisms or their environment and to integrate the DNA into their genome. Horizontal gene transfer has been discussed as a risk of gene escape into the environment without any control. During evolution, horizontal gene transfer has taken place, but it is considered to be a very rare event. Still, it cannot be ruled out and in the context of antibiotic-resistance marker genes this possibility has attracted a lot of attention. According to Directive 2001/18/EC antibiotic-resistance marker genes should be phased out for GMOs to be placed on the market until the end of 2004. Of course, alternative marker genes such as those conferring the possibility of metabolising new substrates, have to undergo new risk assessments.

3.5.4
Potential Trait-Specific Environmental Effects

The potential consequences of the general effects discussed above depend mainly on the transgenic trait of the GMO. Up to now, the main traits for GM crops are herbicide tolerance and pest resistance.

Herbicide tolerance genes confer tolerance to broad spectrum herbicides like glyphosate (Round-Up) or glufosinate (Basta). This trait represented 75% of all GM crops planted commercially in the year 2002. Possible trait-specific environmental effects, apart from the ones discussed in the previous section, are mainly due to the application of the respective herbicide and changes in crop management. Glyphosate and glufosinate are said to be more environmentally friendly than other herbicides in use. Easier and less applications might lead to less pollution of soil and ground water. The possibility of a later application during cultivation could lead to a better soil coverage with plants (weeds) and less erosion. On the other hand a permanent use could reduce biodiversity of weeds and related animals considerably.

In 2002, 17% of commercially planted GM crops world-wide were insect-resistant through the expression of a toxin from the soil bacterium *Bacillus thuringiensis* (Bt) (see also Sect. 2.2). The Bt toxin has been used for many years as a spray in organic agriculture. Out-crossing of Bt-crops resulting in certain advantage for Bt-producing weeds is a potential negative effect. Of greater concern are the unintended effects of Bt plants. This issue has been widely discussed in the context of assumed damage to larvae of the Monarch butterfly in the USA after being fed pollen of Bt-maize in a laboratory setting. Adverse effects could not be confirmed by field trials. Soil organisms might come into contact with the Bt toxin, as it is exuded via the plant roots. The effect on the soil ecosystem is still unclear.

Another important issue is the development of resistance mechanisms against the toxin by the targeted pests. This is a normal process, taking place for conventional synthetic pesticides after approximately 10 years. Development of insect resistance is therefore assumed, which would also render the Bt toxin useless for organic agriculture. The application of certain risk management strategies, with refuge areas where non-Bt-plants are grown to delay the development of resistance, is requested in the USA.

3.5.5
Potential Effects on Human Health

Food consisting of, or derived from, GMOs is tested for potential negative effects on human health according to Regulation (EC) 1829/2003. The assessment includes tests for toxic effects, allergenicity and unfavourable changes in nutrient composition. Not only genetic modification but also plant breeding in general could potentially lead to unexpected or unintended changes in concentration of toxic substances, anti-nutrients or nutrient composition. However, conventional food is not subject to similar examinations.

A starting point for the safety evaluation of GM foods is the application of the concept of substantial equivalence (see also Sect. 4.2.5). This concept was first formulated by OECD in 1993 as a guiding tool and has been developed further since then. Meanwhile it has been internationally accepted, although criticised as being too general and poorly defined. In the EU, with the introduction of Regulation (EC) 1829/2003, the concept has been abandoned. Substantial equivalence is based on the comparison of the GM crop with the appropriate conventional counterpart (considered to be safe on the basis of long experience of use) with respect to phenotype, agronomic characteristics and food composition (key nutrients, antinutrients, toxicants typical of the plant). Three scenarios are distinguished:

1. The GM food or plant is substantially equivalent to its conventional counterpart and is thus considered to be as safe as this conventional counterpart. This is the case when the end product does not contain the newly introduced protein, e.g. sugar from GM sugar beets, or the newly introduced protein has been part of human diet before. No further safety testing would be necessary.
2. The GM food or plant is substantially equivalent except for the inserted trait e.g. the Bt-protein from GM maize. The safety tests would apply only to the newly introduced protein.
3. The GM food or plant is not equivalent to its conventional counterpart. This would be the case for oil from oilseed rape with changed oil composition. In this case the whole plant or food would be subject to safety assessment.

An analysis of key components is carried out to compare GM plants or food with conventional counterparts. The OECD has compiled so called Consensus Documents, the minimal key components of specific crops that should be checked for comparing GM and non-GM crops. Consensus Documents are available for potato, sugar beet, soybean and low erucic acid oilseed rape. International harmonisation is considered necessary to prevent trade barriers. In July 2003 the Codex Alimentarius Commission adopted the "Principles for the risk analysis of foods derived from biotechnology".

Toxicology assessments are not considered to pose any problems with highly purified substances but are more difficult with whole foods. Many conventional crops produce low levels of known toxic substances (e.g. lectins in beans, solanine in potatoes, erucic acid in rapeseed) or antinutrients (e.g. trypsin proteases inhibitors interfering with protein digestion, phytic acid binding minerals). These substances are present at levels significant to human health but are inactivated by food processing, e.g. cooking.

It is difficult to assess the potential for allergenicity. Until today there have been no methods that allow the identification of new proteins as allergenic. Indirect methods are used, based on general characteristics of known allergens as such as typically large protein size, exceptional stability, amino acid sequence homology to known allergens and the quantity of the respective protein in the crop (generally above 1%). Of the huge number of proteins in food, only very few are allergens. Known allergens are found in milk, eggs, peanuts, tree nuts, soybean, fish, crustaceans and wheat. Currently, only in one case has a transgenic protein been shown to be allergenic. A protein from Brazil nut, which was transferred to soybean to enhance the nutritive value for feed purposes, turned out to be a major allergen. This GM soybean has never been marketed. Starlink maize is another GM crop for which potential allergenicity of the newly introduced protein has been discussed. This GM maize contains the Bt protein Cry9C, which could be a potential allergen because it shows some of the general features of allergenic proteins, e.g. molecular weight and relative resistance to gastric proteolytic degradation as well as to heat and acid treatment. For this reason Starlink maize was only authorised to be used for feed in the U.S. However, Starlink maize has been detected in small amounts in maize food products, which put in question the segregation systems in place. Some consumers reported allergic reactions after consumption of maize products, but a connection to Starlink and thus to the Cry9C protein has not been found by U.S. Centers for Disease Control and Prevention (CDC). However, due to some shortcomings in carrying out the investigation, the question of whether or not Cry9C is an allergen still cannot be answered with absolute certainty.

3.5.6
Scientific Uncertainty

In many cases of potential environmental or health risks, the scientific knowledge base is not good enough to assess potential risks in a quantitative way and with sufficient certainty. Profound understanding of complex ecological systems is lacking as well as knowledge to predict the long-term effects of novel food in the diet on the health status. However, it is important to be aware of the fact that this is not only true for GM crops and GM food but also for new varieties of conventional crops and novel exotic foods that have not been consumed in Europe before. It should also be noted that GM crops and food are examined to a much higher extent than any other conventional crop or food.

As quantitative risk assessment is not possible in many cases, a qualitative evaluation system has been developed. The magnitude of potential consequences can be described as negligible, low, moderate and severe. Also the likelihood that these consequences will come into effect can be assigned as negligible, low, moderate or high. The risk must then be assessed by combining the likelihood with the magnitude of consequences. For example, a high magnitude of consequences of an adverse effect combined with a low likelihood of the adverse effect being realised could result in a moderate risk. The final evaluation depends on the specific GMO and needs to be considered on a case-by-case basis.

3.5.7
Risk Management

If the risk assessment identified a risk, a risk management strategy may be developed to minimise or mitigate it (see also Sect. 4.5). A 100% safety or 0% risk is not achievable as a result of risk assessment, therefore uncertainty is an unavoidable part of risk assessment and risk management. Risk management measures could include:
- Confinement strategies, e.g. certain GM crops are only allowed to be grown in greenhouses.
- Restricted use, e.g. the growth of GM crops could be restricted to certain geographical areas.

- Monitoring following experimental release of GM crops or commercialisation of GM crops or GM food. Monitoring can be used to identify predicted or unforeseen effects.
- Guidelines and technical support, e.g. introduction of refuge areas to minimise resistance development of pests or advice for good agricultural practices as such as crop rotation and weed control to avoid weediness of GM crops and GM volunteer plants.
- Record keeping (the use of documentation), e.g. as foreseen in Regulation (EC) 1830/2003 on traceability of GM crops and food as an important part of risk management.

In addition, the design of GM crops could be changed towards male sterile varieties or to the production of sterile seeds (e.g. terminator technology). The latter is especially controversial as the production of sterile seeds will prevent farmers from saving seeds, forcing them to buy new seeds every year.

3.5.8
The Precautionary Principle as Part of Risk Management

Very often scientific data is not available or is insufficient to assess a possible risk in relation to a GM crop in a significant manner (see also Sect. 4.2). Several questions are not addressed due to lack of data on fundamental biological phenomena as such as out-crossing behaviour in oilseed rape or the effects of GM crops on the soil ecosystem. Scientific uncertainty in risk assessment leads to the question of how to deal with risks that cannot be sufficiently quantified. The precautionary principle was introduced at the 1992 Rio Conference on the Environment and Development in Article 15 of the Rio Declaration:

> In order to protect the environment the precautionary approach shall be widely applied by States according to their capability. Where there are threats of serious or irreversible damage, lack of full scientific certainty shall not be pursued as a reason for postponing cost-effective measures to prevent environmental degradation.

The precautionary principle is also included amongst the other international treatise and declarations, and referred to in Directive 2001/18/EC. However, the application of the precautionary principle is not clearly defined and harmonised and gives rise to different interpretations. Generally, the precautionary principle encompasses a forward-looking approach, which includes the prevention of damage, and has a cost-benefit analysis of action or lack of action and the ratio of this response refers to the cost-effectiveness of the action. The application of the precautionary principle should be non-discriminatory and consistent, i.e. comparable situations should not be treated differently and measures should be consistent with measures adopted under similar circumstances. Measures taken have to be reviewed as new scientific developments evolve.

3.5.9
Risk Communication

Risk communication to stakeholders is a key area of risk analysis. The expression of each risk assessment should be unambiguous, transparent and relevant. Key rules, identified by the Scientific Steering Committee (SSC) of the European Commission include:

- Completeness of information
- Public access to documentation
- Transparency of discussions and motivations
- Frank acknowledgement of the various positions and contrasting view, including speculations
- Clarity in wording and accuracy in use of specific expressions
- Recognition of different interests and stakeholders
- Recognition of social, cultural and ethical issues

Awareness of risk perception is another important factor in communicating risk. Risk perception of experts and the general public might differ considerably, because personal opinions are formed by information from different sources and integrated with personal experiences. Among the factors influencing public perception of risk are, for example, the extent to which the risk is voluntary, controllability of the risk and the novelty of the risk form. The SSC suggests expressing conclusions of risk assessment in a more user-acceptable manner by putting them into some form of context, e.g. through risk ranking by comparing risk assessments of different, but related, sources of risk, the risk of possible replacements and by using risk benefit analysis.

3.5.10
Information Sources

Codex Alimentarius (2003) Principles for the risk analysis of foods derived from modern biotechnology. ftp://ftp.fao.org/codex/standard/en/CodexTextsBiotechFoods.pdf
Custers R (ed) (2001) Safety of genetically engineered crops. The report gives a state-of-the-art review on different biological risks of genetically engineered crops. Obtainable from VIB, www.vib.be
EC Scientific Steering Committee (2000) First report on the harmonisation of risk assessment procedures. http://europa.eu.int/comm/food/fs/sc/ssc/out83_en.pdf
European Parliament and Council (1990) Council Directive 90/220/EEC of 23 April 1990 on the deliberate release into the environment of genetically modified organisms. Off J Europ Commun L117:15
European Parliament and Council (1997) Regulation (EC) No 258/97 of the European Parliament and of the Council of 27 January 1997 concerning novel foods and novel food ingredients. Off J Europ Commun L43:1
European Parliament and Council (2001) Directive 2001/18/EC on the deliberate release into the environment of genetically modified organisms. Off J Europ Commun L106:1 http://europa.eu.int/eur-lex/en/search_lif_simple.html
European Parliament and Council (2003) Regulation (EC) No 1829/2003 of the European Parliament and of the Council of 22 September 2003 on genetically modified food and feed. Off J Europ Commun L268:1

European Parliament and Council (2003) Regulation (EC) 1830/2003 Concerning the traceability and labelling of genetically modified organisms and the traceability of food and feed products produced from genetically modified organisms and amending Directive 2001/18/EC. Off J Europ Commun L268:24

König et al (2004) Genetically modified crops in the EU: food safety assessment, regulation, and public concern. Office for Official Publications of the European Communities

Kuiper et al (2001) Assessment of the food safety issues related to genetically modified foods. Plant J 27(6):503–528

OECD (2000) Report of the task force for the safety of novel foods and feeds. Provides an overview on scientific issues and current approaches to food safety assessment. http://www.olis. oecd.org/olis/2000doc.nsf/LinkTo/C(2000)86-ADD1

OECD (2000a) Report of the working group on harmonisation of regulatory oversight in biotechnology. This report is complementary to the report of the task force for the safety of novel foods and feeds (see above). It focuses on the environmental safety implications of the use of products of modern biotechnology. http://www.olis.oecd.org/olis/2000doc.nsf/ LinkTo/C(2000)86-ADD2

OECD Consensus documents for the work on the safety of novel foods and feeds. These documents provide information on compositional considerations for new (GM) varieties of several food and feed crops. http://www.oecd.org/EN/document/0,EN-document-0-nodi-rectorate-no-27–24778–0,00.html

3.6
Monitoring of GMOs

Paul Pechan, Ervin Balazs

Plant biotechnology holds the promise of becoming an increasingly valuable tool in the efforts to improve our heath and achieve sustainable solutions for agriculture and the environment. Improved vaccines, increased food production and more effective waste treatment of polluted lands are but some of the results we may expect (see Sects. 5.1–5.3). However, plant biotechnology may create undesirable side effects. In order to reduce these risks and at the same time fully exploit the potential of this technology, a number of actions need to be taken. The first is the creation and implementation of rules and regulations to govern the application and trade of plant biotechnology products and second, enforcement of these rules through risk assessment, risk monitoring and transparent management. For regulatory related issues please see Sects. 3.3, 3.4 and for risk assessment and management see Sect. 3.5. This section concentrates on the issue of monitoring of GM crops, especially as it relates to their detection.

3.6.1
Why the Need for Monitoring of GM Products

The rapid increase in the commercial scale of transgenic plant in the world from 1.6 million ha in 1996 up to more than 80 million ha today indicates the increasing importance of GM crops worldwide. Public attitude towards GM products varies from total rejection to full acceptance. The many and complex reasons for such varied attitudes are dealt with in Sects. 4.3–4.5. In order to address the societal and environmental concerns, EU legislators have agreed on the general principles of traceability and labelling of GM products to give consumers choice and for ensuring tractability of GM products throughout the entire production and distribution chain. An important part of these new requirements is the monitoring of GMOs.

The requirements for monitoring of GMOs are detailed in Directive 2001/18. The envisaged monitoring plan should be case-specific and used to identify the occurrence of adverse effects on human health and the environment that were not anticipated in the initial risk assessment. The general monitoring has to establish a routine surveillance practice, which includes the regular monitoring of agricultural practice including its phytosanitary and veterinary regimes and medical products. As both plant quarantine and veterinary inspections have internationally recognised control systems their adjustment to include the surveillance of GMOs is also envisaged in some countries.

According to the EC Directive 2001/18, if a notification for deliberate release in a member state is filed, it must include a monitoring plan, accompanied by relevant methodology along with the post-release monitoring. When a GM plant is considered for placing on the market, its monitoring plan is confined for a 10-year period (the time the product is allowed to be marketed under the new regulations). Under the Directive's Article 20 (governing the monitoring and handling of new information) the notifier is responsible for monitoring and reporting to the Commission and the competent authorities of the member states. The competent authorities have the opportunity to communicate with the Commission on new information about the risks the GMO poses to human health and or the environment and thus lodge reasonable objections to further placing on the market of the particular GMO. Member states have the opportunity of provisionally restricting or prohibiting the sale or use of the particular GMO in their sovereign territories if new scientific findings based on the monitoring data have an impact on environmental risk assessment or the potential risks to human health or the environment. At present, such decisions rely to a large degree on our ability to properly monitor each and every GM product placed or to be placed onto the market.

3.6.2
What Needs to be Monitored and How

Monitoring should be seen as part of the decision-making process that also includes risk assessment and risk management. As a general rule, risk assessment addresses product development prior to its eventual placement onto the market (e.g. for field trials or market introduction), while monitoring concentrates on events both prior to and after the product has been authorised for specific use. Monitoring can be seen in specific cases as a part of continuing risk assessment to identify previously unknown or unintended hazards and risks. It may be imposed also as part of precautionary actions. Monitoring can be categorised according to whether we wish to concentrate on the impact of GMOs onto the environment, or the impact on the food industry and consumption, especially since this relates to animal and human welfare.

Due to the different natures of the modified organisms and the introduced traits, all cases should be considered individually on a case by case basis. Case-specific surveillance should be interdisciplinary and carried out over a sufficient timescale to detect any unanticipated delayed or longer term, direct and indirect, health and environmental effects of GMOs. Discussion an the type of monitoring to be carried out is beyond the scape of this section as it will need to be tailored to individual cases. Nevertheless, the first and most important step is common to all monitoring activities: the ability to detect GMOs. Indeed, the current EU labelling and traceability regulatory requirements for GMOs will put an increased focus on monitoring activities in uncontained situations to detect and analyse GM materials that have already been authorised and released for human or animal consumption. This will require that GMOs can be detected at any point within the food chain, from the farm to the market.

A number of technical challenges exist in ensuring the reliable detection and evaluation of GMOs. The challenges can be divided into three groups:

1. Handling and sampling methods, including those needed for identity preservation
2. Detection, identification and quantification methods
3. Availability of reference material

3.6.2.1
Handling and Sampling Methods

Concentrated efforts will be needed to ensure that GM material can be traced, using appropriate sampling procedures, throughout the food chain. Sampling needs to be carried out at the following points:

- Seed suppliers
  Plant breeders will need to assure purity and identity of supplied plant material, to ensure that GM materials can be traced back to their original sources (the so-called material identity preservation). Point of origin sampling and certification will be crucial.
- Farm level
  Farmers will need to keep planting and harvesting equipment clean to avoid cross contamination. They will need to assure there is no cross-pollination between GM and non-GM plants and storage facilities will need to be kept segregated.
- Transport and further storage level
  Random samples will need to be taken to ascertain sample purity and all equipment and storage facilities will need to be kept segregated, or at least clean, to assure there is no cross contamination.
- Processing and distribution
  Each component of the final product will need to be labelled so that its origin could be traced.

Sampling methods are the key to obtaining meaningful qualitative and quantitative results on GM content of food products and any subsequent safety tests. Statistical analysis is an important element of designing appropriate sampling methodologies. Where and how samples are taken as well as size of the sample is critical to a final test result. All samples submitted for testing should be representative of the batch tested. The less uniform the contamination, the higher is the probability of false negative detection. This is crucial for commodity trade as any false results can lead to expensive recalls.

3.6.2.2
Detection and Identification of GMOs

GMOs contain one or more additional characteristics, such as changed protein, sugar or secondary metabolite levels. Genetic modification involves insertion of a foreign piece of DNA into the genome of the organism to be modified. Such foreign sequences can be detected both at the DNA or protein level. In both cases, quantitative and qualitative methods are available, although with different sensitivities. While detection and identification of GM raw material on the farm is relatively easy, in processed food the detection became more and more difficult, indeed in some cases almost impossible. Europe tends to use DNA detection methods and USA relies primarily on the identification of the expressed gene product, i.e. its protein.

Food containing or derived from GMOs needs to be labelled as such. In addition, unintended contamination of non-GM products with GMOs at a level higher than 0.9% requires such non-GM products to be labelled as containing GMOs. It needs to be emphasised that agreed-upon levels of detection (ie 0.9%) have nothing to do with the safety of the product (see also Sect. 3.3). A product with 50% or 100% GMO content is just as safe to eat as a product with no GMO content. If a product is not safe it will not be allowed onto the market regardless of whether it has 0%, 50%, or 100% GMO content. The agreed-upon level of labelling a product GMO or non-GMO has to do with the advances of detection technologies. It has been recommended that tolerance for GMO presence in food products should in the future be based on agreed-upon contamination levels in the supply chain, not on the technological developments of detection sensitivities.

3.6.2.2.1
DNA Analytical Methods

DNA-based detection methods are primarily based on multiplying a specific (for example genetically modified) DNA with the polymerase chain reaction (PCR) technique. Two short pieces of synthetic DNA (called primers) are needed, each complementary to one end of the DNA to be multiplied. During the reaction, copies of the target DNA sequence are made and subsequently visualised. No copy is detected if the target DNA is not present. It is possible to detect DNA in fresh plant tissue, but also in highly processed foods like cakes or chocolate. The PCR results can be quantified giving an estimation of the amount of GM component in the sample tested.

Advantages of the method are: it quantifies molecules of interest (expressed on genomic equivalents basis); it allows GMO content quantification of ingredients in virtually all foods on the market today; the quality of sample preparation is not very important; it has high sensitivity.

Disadvantages of the method are: it does not indicate whether the introduced DNA (gene) is active; it requires skilled technicians.

3.6.2.2.2
Protein Analysis

The method is based on detecting the presence of specific proteins (antigens) with antibodies, and on enzyme assays that detect the activity of a specific protein.Recently a very easy and simple test was developed called lateral flow strips, in which colour-dyed antibodies are fixed to the nitro-cellulose filter are dipped into the extract of the plant tissue bearing the transgenic proteins. The actual reaction time is less than 10 min and allows economical and fast visual evaluation of the results. In food industry, enzyme-linked immunoassays such as ELISA are well-accepted technologies for detection of food contamination.

Advantages of the method are: it indicates whether the new gene is active and to what extent in the recipient organism as indicated by the detected protein, sensitivity for the specific questions to be answered, quantitative (expressed on weight/weight basis), does not need special training or new sophisticated laboratory equipment.

Disadvantages of the method are: quality of the extracted material is important, cannot be used efficiently on processed food, if the protein to be analysed changes in structure it may not be detected, proteins are much more easily degraded than DNA making them more difficult to handle and giving possible false negative results.

3.6.2.3
Reference Material

Detection and analysis of GM samples is useful only if both the positive (certified reference material) and negative controls are available for comparison with the analysed GM samples. These comparisons will greatly increase the confidence and ability to make meaningful conclusions about the analysed samples.

3.6.2.3.1
Certified Reference Material

During the last few years several companies and state agencies have developed reference materials and different PCR systems to standardise their activity for harmonisation of the monitoring methods. Certified GM and non-GM material that is well characterized and of predictable quality is needed to allow laboratories across Europe to calibrate their equipment and procedures. In addition, knowledge of the DNA sequences inserted into the donor material need to be available in order to detect GMOs. Consequently, the new EU legislatures require the GMO applicants to provide sufficient information on the detection methods for each GMO to be authorized for use in the EU.

In recent validation studies at ISPRA Joint Research Centre of the EU, almost 30 laboratories from 13 countries used the same PCR primers to correctly identify transgene from soybean and corn (which contained 2% of foreign materials). The detection was more difficult in the case of corn, which was attributed to its larger genome. This type of extensive validation is necessary to be certain that the observed results are correct. The validation should be performed by independent laboratories using internationally accepted standard methods.

3.6.2.3.2
Need for Valid Comparisons

Besides detecting the presence and levels of expression of transgenes, additional monitoring needs to be carried out to evaluate their safety, for example to ascertain the possible effects of the transgene on the target organism or the environment. This can be, for example, effects on plant metabolism or the population ecology. The results need to be compared to the impact that organic and conventional farming may have in similar circumstances. Such data are needed to make meaningful comparisons between GM and non-GM counterparts. However, our baseline knowledge on current agricultural practice, including organic and conventional farming, is incomplete and indeed may be less than what we now know about GM crops. If there are serious gaps in this knowledge, it follows that monitoring, including gene detection, has to be extended to non-GM crops to gain comparative data.

3.6.3
Future Trends

The future developments in the monitoring of GMOs will depend primarily on three issues:

1. Advancements in the detection technologies
2. Improvements in the baseline knowledge
3. Trends in international agreements in as far as labelling and monitoring of GMOs is concerned

Further developments in PCR technology may lead to lowering of the 0.9% detection and labelling threshold. Already today, it is possible, although not consistently, to detect GMOs when representing only 0.1% of the product being tested. Improvements in baseline knowledge will require that more research money is spent in characterising and assessing non-GM counterparts. Finally, international trade requirements, public opinion and advances in scientific knowledge will likely influence the scope and type of monitoring activities. Internationally agreed-upon traceability and labelling requirements may reduce the costs of monitoring in the long run.

3.6.4
Information Sources

Anklam E, Neumann D (eds)(2002) Method development in relation to regulatory require-
   ments for the detection of GMOs in the food chain. J AOAC Intl 85:751–815
Grasserbauer M, Knowles M, Kuiper HA, Gendre F (eds)(1999) Detection methods for novel
   food derived from genetically modified organisms. Food Control 10(6):339–415
ILSI (2001) Method development in relation to regulatory requirements for the detection of
   GMOs in the food chain. ILSI, Washington, DC
Kjellsson G, Strandberg M (2003) Monitoring and surveillance of genetically modified higher
   plants. Birkhauser, Basel

4 Socio-Economic Considerations

4.1
Implications of the GMO Labelling and Traceability Legislations

Paul Pechan

Consumers are concerned about the safety and usefulness of GM products. In addition, unanswered questions remain about the environmental impacts of GM crops. Partly in response to these developments, the European Union has decided there is a need to label and trace all GMOs from the farm all the way to the market place. While many question the wisdom of this decision, others applaud it.

4.1.1
Background

The number of products from GM crops has increased rapidly over the last two decades. Currently about 70% of processed food may be made from or contain GM ingredients – originating mainly from soybean. Yet this increase has not been reflected in increased sales of GMOs in Europe. Indeed, between 1999 and 2003, the European Union had defacto placed a moratorium on the marketing of GM products. This action in part reflected the publics' concern at that time (see Sects. 4.3 and 4.4). Effective negative campaigns, launched by some NGOs in the 1990s, highlighted the "unnatural" origin of biotechnology-based products and raised doubts about the health and environmental safety of the GM products. Other NGOs, such as European consumer organisations, emphasised the link between consumer acceptance of biotechnology and rigorous and transparent control and labelling of GMOs. The attitudes and concerns of the European public, regardless of how they may have been arrived at, have been reflected in demands for being given the freedom to choose between GM and non-GM products. Partially in response to these concerns and demands, the European Union has enacted several directives to regulate the use of individual GMOs: the most important are arguably the labelling and traceability legislations. The legislative package establishes a system to trace and label GM seeds, crops and food products from the farm all the way to the local market (see Sect. 3.3). The legislation is primarily aimed at increasing the public confidence in the regulatory and enforcement agencies in Europe as well as improving the reliability, efficiency and transparency of the decision making process. In addition, traceability is viewed as part of the cautious approach to health and environmental safety as advocated by the European Commission's Precautionary Principle (see Sect. 4.2). In all cases, the products or processes allowed on the market are deemed to be safe for human or environmental use.

4.1.2
Basic Arguments For and Against the GMO Labelling and Traceability Legislation

A number of different views have been put forward to either argue for or against GMO labelling and traceability legislations. It is important to note that both absence or presence of food labelling can be viewed as misleading communication. When a product is labelled information may be left out or be worded incorrectly, leading to the wrong conclusions on the part of the reader. Cultural differences, knowledge and education have an impact on label interpretation.

Argumentation against labelling and traceability relies on the following main points:

- GM crops and food that are allowed onto the market must be safe, thus there is no need for labelling and traceability. Concerns have been raised that food labels, stating that the food is composed from or contains GM ingredients, may mislead consumers into thinking that it is a warning label. This is misleading. Other health-related information would be more informative and beneficial to the consumers, such as country of origin labelling and whether farming practices and food processing follows agreed-upon international standards.
- Food with a GM label on it does not provide any useful health information to the consumer. GM food is just as safe as non-GM food products. GM food products are thus being discriminated against.
- Food labelling and traceability are impractical. As food ingredients are provided from different suppliers, farmers, food processors and manufacturers from different regions would be required to keep track of all the GM ingredients in the food and whether the GM components are above or below certain allowable levels.
- GMO labelling will cost a lot of money. This money can be used more effectively to improve the safety of non-GM food where, to date, all the proven serious health related problems have been reported. Thus spending money on labelling GM crops is unethical.
- Labelling and traceability may hinder planting of GM crops (see also Sects. 3.5 and 4.1).
- If GMO products are substantially equivalent to the non-GM counterparts, there should be no need to label them. Substantial equivalence can be defined as indication that the composition, nutritional value, or intended use of GM food has not been altered.
- Developing countries that decide to grow GM crops because of higher yields and lower cost inputs would suffer because of the additional testing requirements to meet the labelling and traceability requirements. Indeed, traceability legislations may force these countries not to grow GM crops and thus increase their reliance on environmentally damaging pesticides and herbicides. In addition, this may have a negative effect on trade (see also Sect. 4.2).

In conclusion, labelling and traceability of GMOs is not needed. It has been introduced primarily to address the public concerns that have been fuelled by negative publicity spread by a few environmental organisations. It will cost a lot of taxpayer money while providing minimum benefits to the public. There are real health and environmental issues that require legislative attention. GMOs are not one of them.

Arguments for GMO labelling and traceability rely primarily on freedom of choice and on the precautionary principle, which advocates caution with new technologies Both provide confidence in safety regulations and public institutions.

- Labelling and traceability are needed because the farmers, food processors and end users should have the opportunity to segregate GM from non-GM food as necessary and have a choice of what food to buy.
- Consumer organisations want labelling of GMOs, not because of real health or safety issues, but because the public debate in Europe has made it impossible not to label GMOs. The issue has become a question of preferences and choices based on personal values as well as on other complex issues such as globalisation and international trade. These desires have to be respected and appropriate label information provided so that consumers can make an informed choice about their buying preferences. This may be important; for example to people of certain religions.
- Although evidence to date indicates that there have been no illnesses related to GM food or environmental damage, the traceability regulation ensures that if something does go wrong, the GMO in question can be rapidly withdrawn from the market.
- Labelling and traceability will preserve the purity of the source or nature of material. This point is very important to the European organic food industry.

In conclusion, GMOs should be labelled and traced. It is a new technology with new products and the public should have the choice and time to accept or reject it on their own terms. Traceability is simply a prudent step to assure that if, for whatever reason, something does go wrong with this new technology, the product in question can be quickly identified and withdrawn from the marketplace. As traceability and labelling are well-established concepts, the costs are not going to be high. In addition, the paper trail established on the basis of the traceability legislation will assure that the monitoring and testing costs are kept to a minimum.

4.1.3
GMO Labelling and Traceability Legislations Outside of Europe

Because of the rapid changes in legislations, the reader is advised to consult the relevant internet sites for the latest information.

4.1.3.1
United States

In 1966, the US Congress passed the Fair Packaging and Labelling Act requiring that all consumer products be honestly and informatively labelled. However, decisions about GMO labelling follow a different path to that of Europe. The USA requires mandatory labelling of GM foods in cases where there are health or nutritional concerns. The legislation concentrates on the product rather than the process of creating the product, as advocated in the EU. In January 2001, the USDA released a new guidance for industry for voluntary labelling of foods that use bioengineering (see also Sect. 3.3).

4.1.3.2
Canada

GM foods must be labelled in cases where the modification has resulted in potential health or safety concerns such as a nutritional or compositional change or an increase in potential allergenicity. In such cases, the labelling must indicate the nature of the change. The method by which that change was achieved (i.e. through genetic modification) is not required. This is consistent with Canada's regulatory approach, which focuses on the properties of the product rather on the process through which the product is developed.

4.1.3.3
Australia/New Zealand

Mandatory labelling requirements have been announced effective December 2001. Labelling is required in cases where foods have altered characteristics, such as changed nutritional values, or when foods contain novel DNA or protein as a result of genetic modification.

4.1.3.4
Japan

In 2001, Japan's Health, Labour and Welfare Ministry issued guidelines for the mandatory labelling of food products containing GMOs. The labelling regulations applies to GM ingredients if the ingredient is one of the top three food ingredients by weight and composes at least 5% of the total weight of the product. Labelling is not required on packages of less than 30 cm^2. Food products containing up to 5% of approved biotechnology crops such as corn and soybeans do not need to be labelled as containing GMOs.

4.1.4
Impact of the GMO Labelling and Traceability Legislations

4.1.4.1
What Products to Label

The GMO labelling and traceability legislations require that all products made from, containing, or composed of GMOs must be labelled as GM products. Although the application of the legislation appears to be straightforward for GM crops, there is a discrepancy in applying the rules to processed food and indirect use of GM products. Why is it that cheese makers that use GM bacteria in cheese processing (perhaps also fed on a starch derived from GM soybean) do not need to label the cheese as a GM product, while an oil that contains no traces of GM material must be labelled? Why are animals that are fed on GM soybean excluded from labelling and traceability requirements? If the purpose of the legislations is to give choice and security, it would be consistent to label all the above products as containing or being derived from GMOs. Moreover, why choose 0.9% as the threshold for labelling in the case of unintended or unavoidable admixtures? Why not 2 or 5%? These points illustrate that the labelling part of the legislation is not based on science but rather on a political compromise, attempting to satisfy the needs and demands of the key stakeholders.

What is evident is that the labelling legislation was needed to restore consumer confidence in the EU decision making process. It reflects the concerns of the public. However, rules alone cannot restore this confidence. Effective communication and dialogue based on trust is also required (see Sects. 4.3 and 4.4). Moreover, misleading campaigns must be effectively counteracted in order to prevent manipulation of public opinion (see Sect. 4.4). Otherwise, legislation may reflect the views of a few groups with effective public relations campaigns, rather than those of the public.

As a footnote to the labelling debate, it is surprising that the industry, once it became clear that there would be a labelling legislation, did not insist on labelling of all food products that came into contact with GMOs. In that scenario, so many products would have had to be labelled, that within a short time, the public would have become "acclimatised" to GMOs. Even in the present situation, a justified concern of NGOs opposing GMOs is that consumers are unlikely to read food labels. This explains the continuing pressure of some NGOs on supermarket chains not to allow GM products into their stores, and why some food biotechnology companies in Europe are currently concentrating their publicity efforts in the same area.

4.1.4.2
Impact on Farming Practices

There are three basic farming methods in Europe: conventional, organic and GM. They are trying to co-exist within the confines of the agricultural realities in Europe. The problem for GMOs is that proponents of organic farming have created standards that are incompatible with GM crops. The standards call for zero tolerance of admixture from GM feed or seed. Organic farming emphasises its respect of nature and environmentally friendly technologies. This attitude should encourage consumers agreeing with this philosophy to choose organic products over all other products. GM crops, on the other hand, have been marketed with emphasis on the economic benefits to farmers. Both agricultural systems claim to be safe, both for the human health and the environment.

The coexistence measures should follow the EU 0.9% threshold rule for labelling of unintended admixtures However, the European Commission coexistence guidelines make it possible for member states to decide on grounds that GMO agricultural practices could endanger the existence of organic agriculture, to essentially prevent GMOs from being planted in many regions of Europe. The concerned national or regional authorities could, for example, impose such a decision by invoking the precautionary principle (if the extent of the possible GMO "contamination" is not yet clear). Elements of the precautionary principle are embedded in the various EU directives and regulations (see Sect. 4.2), such as Article 23 in Directive 2001/18 (previously Article 16 of Directive 90/220). In effect, member states are given a green light to take whatever steps needed (including temporary ban on GM crops) to prevent unintended presence of GMOs in organic products if that would lead to economic losses to organic products through endangering existing organic farming purity standards. It is likely that only a few GM crops, such as GM potatoes have any chance of being grown in Europe. However, it should be emphasised that whatever actions are taken at member state level, they cannot be discriminatory.

Moreover, according to the coexistence guidelines, farmers that introduce new production types into a region should bear the responsibility of implementing actions to limit admixture. This also includes continuous monitoring and sharing of best practice information. National liability laws and liability criteria for insurance schemes, as they apply to damage resulting from admixture, need to examined and evaluated. At present, EU rules allow member states to take appropriate measures to avoid unintended presence of GMOs in other products (IP/03/1056). Because GMOs are likely to be the new production type introduced into a region, liability (and thus cost) responsibilities will lie with farmers planting GM crops. So long as uncertainty remains about current national liability laws, it is unlikely that many private farmers will take the risk of growing GM crops. It is expected that the liability laws, combined with the zero tolerance of organic farmers to adventitious presence of GMOs, will prevent EU farmers from growing GM crops in many parts of Europe for some time (see also Sect. 2.5.6). As there is no safety objection to the import of GMOs, some EU states will be in the paradoxical situation of trying to prevent GMOs from being grown locally, while having to allow GM products from abroad into their country.

4.1.4.3
Impact on Costs Associated with Labelling and Traceability of GMOs

Food processors will likely need to pay premium prices for guaranteed non-GM materials and test for the presence of GMOs in all other cases, adding more costs to the final product price. In addition, all involved in the production and resale of GMOs will need to assure that they can provide the relevant authorities with a paper trail showing where they bought the material and how they used it. This may be relatively simple for importing GM seed, but very complex in processed foods where a number of ingredients may have GM origin. Because of the differences between products there are at present no estimates of the additional costs: they may range anywhere between zero and 40% of the product costs. Monitoring requirements will likely constitute the largest component of these costs.

4.1.4.4
International Implications

The European labelling and traceability legislations have partly been the reason why the USA filed a restriction of trade lawsuit with the WTO against the EU. The basic argument states that just because the product is genetically modified it does not need to be labelled as such if it is equivalent and just as safe as its non-GM counterpart (see Sect. 4.2 for more on international trade issues). This dispute may help to be resolved by Codex Alimentarius.

Codex Alimentarius was created in 1962 by Food and Agriculture Organisation (FAO) and the World Health Organisation (WHO), both United Nations organisations with the responsibility to harmonise food laws and agree upon international standards through recommendations and guidelines. The Codex Committee on Food Labelling (CCFL) is now discussing whether to expand current country of origin food labelling to include labelling ingredients of composite foods, including a mandatory process-based labelling of GM foods. There appears to be consensus that labelling is needed for foods derived from modern technology when there are significant changes in composition, nutritional value, or intended use, and it is important to provide such information to consumers. The CCFL has achieved a consensus on the labelling of allergens in foods derived from modern biotechnology, and believes that such provisions provide considerable assistance to and protection for consumers. However, there is currently no consensus among Codex countries about a mandatory process-based labelling of foods derived from modern biotechnology. The actions of CCFL have a direct effect on international trade, where the aim is to reduce trade barriers between countries because Codex Alimentarius standards, guidelines, and recommendations are cited by the WTO as the preferred international measures for facilitating international trade in food.

4.1.5
Information Sources

Codex Alimentarius. Homepage, including minutes of all their meetings, can be found at http://www.codexalimentarius.net/agend.htm
Co-existence. A number of European Commission articles and links on this issue can be obtained under http://www.europa.eu.int/comm/food/risk
European Parliament and Council (2001) Off J Europ Commun L106:1
European Parliament and Council (2003) Off J Europ Commun L268:1
Labelling issues. US views on food labelling can be obtained at http://www.usinfo.state.gov/journals/ites/0502/ijee/foodlabeling.htm#mat
Noussair C, Robin S, Ruffieux B (2002) Do consumers not care about biotech foods or do they just not read the labels? Econ Lett 75(1):47–53, http://www.elsevier.com/homepage/sae/econworld/econbase/ecolet/frame.htm
Public perception of GMOs. See http://www.checkbiotech.org/pdf/pubperc.pdf
Refer also to information sources listed in Sect. 3.3.
Reiss M (2002) Labelling GM foods – the ethical way forward. Nature Biotechnol 20(9):868
USDA. Homepage is a good source of information on GM crop usage http://www.ers.usda.gov/Briefing/biotechnology/chapter1.htm

4.2
Precautionary Principle

Paul Pechan

If there is uncertainty as to the extent of risks to human health and the environment, decision makers may take protective measures without waiting until the seriousness of those risks becomes fully apparent. This simple statement is at the core of what is called the precautionary principle. It has become an important decision-making tool in the European Union. It gives decision makers the responsibility to act in the face of scientific uncertainly about possible harmful effects of a technology, product or its application. It can thus call for additional scrutiny of GMOs that have undergone or are undergoing risk assessment. In applying the precautionary principle, decision makers consider scientific as well as social, cultural and economic implications and dimensions of the potential risk. Precautionary actions, initiated by triggering the precautionary principle, are of temporary nature until scientific information is obtained that can ascertain the likelihood and impact of the risk on human health and/or the environment. There are three stated objections to the precautionary principle:

- first, that it may inhibit innovation,
- second, that it is not based on science and
- third, that it can be an excuse for protectionism that gives advantage to specific interest groups.

Without an international agreement on the meaning and application of the precautionary principle, there is a real possibility that trade wars may erupt in the coming years. The challenge is to draw up appropriate terms of reference. These have three components:

1. It is essential to have, a priori, a common agreement as to what the triggers of the precautionary principle are. For example, it needs to be agreed what constitutes a hazard and possible risk to human health and the environment.
2. Once imposed, precautionary actions need to reduce uncertainty about the risk effects. This will require additional research. Agreements between the stakeholders, likely on case-by-case basis, will need to be reached on the questions to be asked, assessment standards and the procedural rules to gather the results.
3. It is important to define the point when to say that we now know enough to make a permanent rather than a temporary decision about the risk posed by an identified hazard. Ethical rather than socio-economic considerations should be discussed and incorporated, alongside science, into terms of reference.

4.2.1
Introduction

Scientific evaluation of a defined problem is an integral part of the decision-making process. However, what happens when scientists cannot provide statistically reliable data, the data does not exist or scientists themselves disagree on the implications of their findings and the resulting risks? The precautionary principle evolved to help decision makers deal with such uncertain situations. It was felt that decision makers should be in a position to make temporary decisions without taking the risk that, if they wait for more conclusive information, it may be too late to take corrective measures to avoid harm to the society or the environment.

The debates over nuclear power plants in Germany in the 1970s can be taken one of the key events in the development of the precautionary principle concept. It solidified various notions of precautionary actions into a unified concept to protect our society and the environment from an identified hazard that has as yet an unclear level and scope of impact. The principle is therefore intended for use when the risks of a hazard are not yet fully understood. Elements of the precautionary principle have been integrated into the European Union treaty and into the Cartagena Protocol on Biosafety that regulates international movement of living modified organisms (LMOs). Because of inclusion of socio-economic considerations, the precautionary principle is often discussed in relation to broader issues such as globalisation and world trade.

Some examples where precautionary action could be considered in view of insufficient data and possible negative impacts are:

- Biodiversity conservation
- Biotechnology
- Nuclear and chemical waste disposal
- Air pollution (both particulate and chemical)
- CO_2 emission levels
- Control of emerging diseases
- Bioterrorism

What these examples have in common is that their potential impact on our health or the environment may be long term, irreversible, cumulative or high impact. In the 12 "late lessons" (Late lessons from early warnings: the precautionary principle 1896–2000, European Environment Agency 2001), it was concluded that neglect of the precautionary principle can be harmful but also that over-precaution can be expensive in terms of lost opportunities for innovation though preventing or neglecting scientific enquiry and its application.

This section concentrates predominantly on discussing the precautionary principle in relation to GMOs.

4.2.2
Philosophy Behind the Precautionary Principle

All scientists would agree that there are instances of uncertainty in the interpretation of experimental data. One way scientists get around this dilemma is by asking specific questions which, it is hoped, can support or reject a particular hypothesis. However, answers to very specific questions can go only a part way to illuminate complexities of life. The precautionary principle recognises that there are limits to scientific knowledge and that there may be instances when not enough information is available about a suspected harm, the likelihood of its occurrence and scope of its impact on the society and/or the environment. These are the so-called instances of scientific uncertainty. It is under these conditions that the precautionary principle can be invoked.

The precautionary principle is not a scientific approach, it is a tool used by decision makers to make active political decisions for the protection of society and the environment in light of scientific uncertainty. The principle is thus meant to help facilitate management decisions. The precautionary principle forces decision makers to take on the responsibility to anticipate and prevent unacceptable harm to the society or the environment. The precautionary principle and the ensuring precautionary actions should amount to prudent actions on the part of appropriate institutions to safeguard the public and the environment against a hazard that may cause unacceptable harm.

The precautionary principle can be used, for example, to put in place a moratorium or, on case-by-case basis, gradually phase in or re-evaluate technologies, products or processes. It is a temporary measure, reviewed in light of new information. Thus new research, monitoring and risk assessment are an integral part of precautionary action. Only on the basis of new information that clearly identifies the risk levels and scope of an identified harm, can final recommendations be made whether or not to permanently ban, withdraw or allow a product, process or technology onto the market. This management approach has broad applications and implications in the decision-making process, regardless of the specific area of concern.

4.2.3
When and How is the Precautionary Principle Implemented:
Current Practice in Europe

Two main aspects of the precautionary principle that need to be looked at are:

1. When is it invoked?
2. How is it implemented?

4.2.3.1
When to Invoke the Precautionary Principle

Two pre-requisites are necessary for triggering the precautionary principle. The first pre-requisite is the identification of a potential harm, originating from a phenomena, product, or process, that can impact on the health and safety of society and/or the environment. Such harm can emerge for example from an existing or emerging technology. The ability to decide what constitutes a possible harm is dependent on the existing level of health or environmental protection standards of the country or region, subject to the general principles of risk management (i.e. proportionality, consistency and non-discrimination (see also Table 4.1). The second pre-requisite is that the harm cannot be assessed with sufficient certainty, for instance because of insufficiency of data, their inconclusive nature or divergence of scientific opinion. Fulfilment of this pre-condition is dependent on information, or rather lack of information, obtained from existing risk assessment data.

Table 4.1. European Commission guideline for the application of the precautionary principle

Non-discriminatory	Comparable situations should be treated in a similar way
Proportional	Measures to achieve appropriate level of protection. This takes into consideration immediate and long-term risks
Consistency	With measures already adapted in similar circumstances using similar approaches. It thus cannot be arbitrary in nature
Examination of cost/benefits of action and lack of action	Measure must provide an overall advantage with regards to reducing risk to acceptable levels. Examination includes economic and non-economic considerations as well as impact of other options
Examination of scientific developments	Protection measures to be in place as long as data is inadequate, imprecise or inconclusive and as long as risk is too high to be imposed on society

In cases where there is little or no uncertainty about the scientific data and assessment of the level of risk, management decisions will be taken that follow already well-established procedures without the need to invoke the precautionary principle.

The challenge of invoking the precautionary principle is primarily to decide what should be the acceptable levels of health and environmental protection standards (against harm), what is an intolerable level of uncertainty and properely weighing risks and benefits. A consensus on benefits these issues is needed before precautionary actions can be considered internationally. Section 4.2.8 discusses this issue in more detail. Consultations with all the stakcholders need to precede the decision whether to impose the precautionary principle.

4.2.3.2
How to Implement the Precautionary Principle

Once the precautionary principle is triggered, a choice must be made about the appropriate precautionary action. The action has two components.

First, temporary measures need to be imposed. The European Commission points out that imposition of (temporary) precautionary measures implies regulation of subject matter on the basis of protection standards that remain open for discussion. The regulation does not define these standards but sets a number of guiding principles (see Table 4.1). This means that the standards may need to be decided on case-by-case and region-to-region basis. Due to the uncertainty about the risk being faced, the possibility of wrong decisions can also be relatively high. It is thus not surprising that decision makers err on the side of caution.

Second, new data needs to be collected and analysed in order to fill in knowledge gaps. Collection of new data implies reaching a higher level of certainty and knowledge about the possible harm that is based upon new (basic) experiments, risk assessment and monitoring activities.

Risk assessment contains a number of well-defined procedural steps. These are hazard identification, hazard characterisation, appraisal of exposure and risk characterisation. As the precautionary principle has been triggered, this means that the hazard has already been identified and likely characterised. However, all other elements of the risk assessment procedure may need to be re-evaluated and/or new data collected and analysed. For instance, there may be uncertainties about who may be at risk and the likelihood of the hazardous event occurring and/or the level of impact, if it does occur.

Input from science is expected at all levels and throughout the whole process of precautionary actions – from basic research to risk assessment and monitoring. It is unclear when and how socio-economic considerations come into play. It is assumed throughout the process, although economic considerations are considered only after health and environmental safety is assured (see below).

At the end of the process, decision makers must weigh up the risk assessment evidence and decide on a further course of action. When, on the basis of new knowledge generated, an acceptable level of scientific certainty is achieved, management decision can be taken to review the precautionary actions. This allows permanent decisions to be taken on whether to allow or disallow an existing or emerging technology, product or process onto the market. According to the European Commission, the application of the precautionary principle is to be the least trade-restrictive measure taking into account technical and economic feasibility after the health and environmental protection concerns are fully addressed. The aim of applying the precautionary principle is said to prevent harm, not to prevent progress. These are, however, statements of intent. Serious concerns remain about the ability to assure transparency, fairness and speed in a decision-making process that is based on the precautionary principle (see also Sects. 4.2.6 to 4.2.9).

4.2.4
Legal and Political Framework for the Precautionary Principle

The European Union does not have a legal document describing and defining the precautionary principle as it relates to both health and environmental issues. The Maastricht European Union Treaty does not define the precautionary principle but states that precautionary measures should be implemented in all EU policies related to environmental protection. A judgement by the European Court of Justice defined the conditions for the application of precautionary measures to human health Community law (Case C-157/96, National Farmers Union, judgement of 5.5.98, ECR 1998, page I-2211, ground 63).

Because the precautionary principle is triggered when there is uncertainty about the scientific results, the decision whether to invoke the precautionary principle and what measures to take is seen as predominantly a political decision. According to the European Commission, the term political decision means risk management decision taken by the responsible regulatory authorities. That decision is "essentially a political or a societal value judgement to be taken by the responsible regulatory authorities". Further, the European Commission "believes that the diversity and complexity of socio-economic conditions could influence the choice of appropriate level of health or environmental protection to be applied by a national community, but that the concepts of science, scientific uncertainty and risk are objective concepts of general and potentially universal application".

Although competencies for risk management both at the European and national levels are relatively well established, albeit in the need of streamlining, the rules for the governance of the precautionary principle is still very much evolving. The ideas represented by the precautionary principle are being integrated into various EU legislations.

The legal framework for the application of the precautionary principle at an international level is in its infancy. A start has been made by embedding it into the Cartagena Protocol. However, a number of countries, most notably USA, claim that the agreement refers to the precautionary approach, whereas the European Union would claim that the Protocol refers to the same basic rationale of the precautionary principle in its decision-making procedures.

4.2.5
Precautionary Principle Versus Precautionary Approach

The difference between these two terms is very important. Approval of new products in the USA is based on the idea of precautionary approach: precautions are to be taken while developing a product to make sure it is safe. USA sees no need to acknowledge the precautionary principle as an international standard as precaution is already inherent to all science-based risk assessment when dealing with issues of safety.

The European Union precautionary principle differs from the precautionary approach, as advocated by the USA, in two important aspects. First, unlike precautionary approach, the precautionary principle takes current societal values and attitudes openly into consideration and second, defines safety standards differently, by taking a broader view of health and environmental safety that reflects long-term sustainability concerns. In other words, the frame of reference and philosophy between the two concepts differ fundamentally (see also Sect. 4.2.8).

Europe appears to take a long-term "sustainable" strategic view of the world with emphasis on governance, by creating a legal framework to minimise human health and environmental risks. Decision makers are asked to take proactive decisions, on a continuing basis, to prevent disasters. In many respects, this can be seen as a reaction to the inability of the existing European monitoring systems to prevent past food-related disasters, combined with very low public trust in political institutions. The USA is more likely to emphasise economic growth with minimal interference from decision makers, once the rules of the game have been agreed upon. Risk assessment standards for GMOs in the USA are, for example, based on the notion of substantial equivalence. The regulators look at a final product not how the product was derived (as is the case in Europe). The concept of substantial equivalence, as defined by OECD, embodies the idea that existing organisms used as foods, or as a source of food, can be used as the basis for comparison when assessing the safety of human consumption of a food or food component that has been modified or is new. In contrast, Europe looks at GMOs not from the product perspective, rather from the perspective of whether a process of genetic manipulation has taken place. The precautionary approach assumes that it is possible to decide whether to allow a product onto the market.

It is up to the central regulatory agencies to ask for more data if some safety doubts remain. When fully satisfied with the results and the expected standards are reached, a given product can enter the market place (see Sect. 3.3.2). Once allowed onto the market, re-evaluation of a product is triggered by a breach of health/environmental standards or is based on new information provided by the industry.

The differences between and use of terms precautionary approach and precautionary principle are often not very clear. For example, both expressions have been used to justify the European Union approach in dealing with potential risks. The precautionary approach also includes socio-economic interests, but they are hidden within the definitions of applied standards, for example the use of substantial equivalence can be viewed as being industry friendly. Although different in some important aspects, both terms also include overlapping ideas of health and environmental safety.

4.2.6
Application of the Precautionary Principle to GM Crops and Food

There are three main reasons to consider invoking and applying the precautionary principle to GM crops, products and processes:

1. Genetic modification of plants is a new technology with little or no possibility for direct comparison to non-GM crops (partially because such baseline data on non-GM crops do not exist)
2. Genetic modification of plants can be inherited and propagated from generation to generation and thus potentially cause irreversible damage to the environment
3. GM food has, compared to our traditional food, not been on the market for very long and thus may pose some, as yet unknown, long-term risks of significant scale and impact

A number of precautionary legislations have been implemented that relate to GM product safety both to the general public and the environment. Among them are the novel food, deliberate release, transport, traceability and labelling legislatures (see also Sects. 3.3 and 4.1).

4.2.6.1
Health Assessment of GM Products

GM products have been on the market since the mid-1990s and all available scientific evidence suggests that there are no health risks associated with eating the GM food currently on the market (see Sects. 2.1–2.3). Rigorous procedures and review processes are in place to ascertain that they are safe for human consumption (see Sect. 3.5). Consequently, invocation of the precautionary principle for existing GM products already on the market in the USA and those being imported into Europe cannot be justified on health risk criteria as they have been shown to be safe.

The situation is, however, different for the new second and third generation of GM products still in the experimental phase of development (see Sect. 5.1 and

5.2). These products have deliberately changed metabolic pathways and are not substantially equivalent to their non-GM counterparts. Therefore, more precaution and testing is required. Here, the precautionary principle can and should be invoked to test their safety more extensively. To be non-discriminatory, this approach needs to apply to any novel products, whether or not they are of GM or non-GM origin.

4.2.6.2
Deliberate Release of GMOs into the Environment

There is insufficient scientific data to understand the complexities of nature. Moreover, scientists often have different explanations of the available information. Knowledge changes and expands over time. The precautionary principle can be invoked because of these scientific uncertainties. It is thus not surprising that a lot of emphasis has been placed on environmental impact assessment studies of deliberate release of GM crops. The concern about currently applied standards in some countries is that direct and indirect harm to the environment is not sufficiently addressed, that the experimental designs are limited in scope (geographical, temporal or sample size) and that the key terms of reference under which risk managers operate is too narrowly defined. These are based predominantly on comparisons to existing conventional agricultural practices.

The European Directive 2001/18, concerning the deliberate release of GMOs into the environment, can be viewed as the first legislation in which the precautionary principle is translated into precautionary legislation. The legislature is structured to deal with each new GM crop on case-by-case basis using a step-by-step procedure. The legislation facilitates mandatory scientific evaluation by member states for every single GMO release. The case-by-case and step-by-step procedure should allow continuous evaluation and a gradual introduction of GMOs into the environment. Concerns about the legislature are threefold:

1. There may be so many potential risk concerns addressed that no conclusive answer will be forthcoming within a reasonable period of time, in effect blocking the development of new crops by using too broad a frame of reference.
2. Decisions by authorities may be taken to appease unfounded public concerns or specific interest groups (rather than decide on basis of scientific evidence only).
3. Although implied in the precautionary principle, there are no practical provisions for concurrent risk evaluation and comparison of alternative options, such as organic farming.

4.2.6.3
Biosafety

The Convention on Biological Diversity in 2000 included reference to the Catagena Protocol on Biosafety of trans-boundary movement of living modified organisms (LMOs), focusing on LMOs that may have an adverse effect on biological diversity. The Protocol refers to the precautionary principle both as a general principle and as a decision-making tool. Some countries, especially the USA, would argue that the reference to the precautionary principle refers to the idea of precautionary approach, not to any inherent principle. The European Commission argument is that what counts is that the core element of the principle is now already accepted internationally, namely that responsible authorities can and should make decisions in order to achieve the chosen level of protection. The Protocol's decision procedures do include case-by-case risk assessment and decision-making processes and, where scientific uncertainty exists, provisions for banning or restrictions on a LMO.

Some exporting countries would prefer to have the Protocol subordinate to the World Trade Organisation (WTO) rules in case the Protocol would give rise to trade conflicts. Indeed, the Protocol may be in conflict with the Agreement on Sanitary and Phytosanitary Measures (ASP) and the Agreement on Technical Barriers to Trade (TBT) of the WTO as they also deal with use and trade of LMOs. A situation may arise where it may not be clear whether the Protocol or the ASP should be used when an LMO dispute arises. One of the legal arguments is that since the Biosafety Protocol was concluded later than the ASP or TBT agreements, the Cartagena Protocol should be used in international disputes. However, it is generally agreed by the European Union that measures based on the precautionary principle (and thus the Catagena Protocol) should comply with basic trade law principles, such as non-discrimination, least-trade restrictiveness and transparency.

4.2.6.4
Traceability of GMO Products

The traceability legislation in the European Union also takes a precautionary view of GM products. It requires that all GM products can be traced from their source all the way to the end user. This is in order to ensure that, if something goes wrong, the product can be identified, no matter where in the food chain, and appropriate precautionary action taken, including withdrawal from the market. Traceability addresses issues of potential health risks and also risks of possible damage to the environment (see also Sects. 3.3, 3.5, and 4.1).

4.2.6.5
Labelling of GM Products

The labelling and traceability legislations are linked through the definitions of what constitutes a GMO and of acceptable levels of adventitious presence of GMO in a non-GMO plant or product. Identification of a product as a GMO allows for its subsequent tracing. Labelling of GM products in itself is not related to safety issues. If a product is not safe, or if doubt exists about product safety, it would not be allowed onto the market. Labelling should give consumers choice in what they buy and, as such, is not directly related to the precautionary principle which can be applied only in cases of suspected health and environmental concerns.

4.2.7
Implication of the Precautionary Principle for International Trade

Two basic aspects of the precautionary principle will be considered. First, what is the impact of the precautionary principle on the competitiveness of innovative European companies internationally and second, how useful is the principle to reach international trade agreements.

4.2.7.1
Competitiveness of European Science and Innovative Companies

The precautionary principle asks decision makers to take action in case of uncertainty. Uncertainty inherently asks people to be careful. The precautionary principle thus offers decision makers the secure option of "let's pause and wait for more data". There is little risk involved for decision makers to take such a view. Politically, such decisions may even be very popular, resulting in short-term gains because the long-term harm of such a decision is not yet obvious. Indeed, precautionary actions to temporarily stop something are a better guarantee to protect their careers that taking the "risk" to let something proceed. By slowing down and complicating the approval process, new innovative technologies may fail to become competitive as this requires speed, flexibility and the appropriate political climate. A way forward is to place the precautionary principle under the leadership of politically independent risk managers (see also Sect. 4.2.10). Also for science, the precautionary principle impacts the role of the scientist within the society as well as influences whether the focus is on creative (basic research leading to knowledge creation) or applied (reactive) research.

4.2.7.2
Impact of Traceability Requirements on developing countries

Traceability requirements advocate caution in placing GMOs onto the market by making sure that GM products can be traced and withdrawn from the market, whenever a risk is identified. However, the traceability and associated labelling requirements may make it more difficult for developing countries to grow GM crops because of inadequate infrastructure, including lack of trained personnel and testing laboratories to comply with the product identity preservation requirements as outlined in Directive 1830/2003. Rather than risk trading problems with the EU, developing countries may decide not to grow GM crops. Perhaps the main difficulty is expected from possible unintended contamination of non-GM crops or seeds with GM material, either in the field, during transit or in storage. The usual practice of farmers of saving seed for sowing in the following season may add another source of mixing GM and non-GM seeds. Although developing countries should always compare alternative solutions to their problems, a priori exclusion of GMOs, as one of the possible solutions, may rob the country or region from effectively addressing, for example, specific insect or weed problems. The European Union or the FAO should provide help in building up the necessary infrastructure to allow developing countries exercise a true choice in agricultural practices.

4.2.7.3
Precautionary Principle and International Trade Agreements

In the context of international trade, every country has the right to apply guidelines that it deems to be appropriate. Every country is autonomous in their decisions. However, all decisions have to be justified in the light of existing scientific data. In effect, a country needs to convince all others that its decision is fair and non-discriminatory. This is true also in cases of precautionary actions.

The basic idea of the precautionary principle is that it is triggered when a possible risk to human health or the environment is identified. This assumes that there is a consensus on the meaning of hazard, the levels of risk we are willing to tolerate and what actions we should take. Within closely co-operating regions, such as the EU or USA and Canada, this is possible. It is much harder to reach such a consensus between these regions or on a world-wide scale.

Different countries may have different definitions of hazard, risk and harm and different standards of what level of safety is acceptable in the light of a possible harm arising from an identified hazard (see also Sect. 4.2.6). This is also influenced by societal values and attitudes, as illustrated by the difference attitudes between the EU and USA concerning the precautionary principle and precautionary action. This is recognised by the precautionary principle. It postulates that risk management decisions are ultimately political, not scientific, decisions. Decision makers do not only consider scientific evidence but also social, cultural and economic dimensions of the region. However, when such considerations are made after risk assessment is completed, no matter how good the intentions, questions will inevitably be asked about the subjectivity and vested interests in the decisions reached. The precautionary approach, as practised in the USA, is also not free from socio-economic pressures. The use of substantial equivalence in risk analysis of GMOs is as much a political and economic statement as it is a valuable, albeit imperfect, concept for risk assessment. Combined, the precautionary principle and precautionary approach may create contradictory needs and objectives and ultimately lead to international conflicts.

In the case of GMOs, the precautionary principle advocates the combination of case-by-case evaluation and flexible standards for ongoing deliberations. The difficulty and challenge is to set up international standards and regulations for GMOs based on the precautionary principle. The decision-making bodies need to agree on: (a) the conditions to trigger precautionary actions; (b) the key questions to be asked and answered and the methods that should be used to clarify that GMOs pose no health and environmental risks, and; (c) the acceptable level of uncertainty in the final decision (see Sect. 4.2.8 for detailed discussion). Disagreements between countries arise due to different definition of these standards, how they are applied to cases of uncertainty and risk reduction (frames of reference) and ultimately whether or to what degree the decisions should be politically or scientifically based.

One possible scenario under the precautionary principle is that if socio-economic aspects take precedence over scientific reasoning, USA or Europe could justify their opposing positions on the basis of regional needs. The end result is that products allowed on the market according to one set of standards, may be rejected on the basis of another. Emphasis on bio ethical rather than socio-economic aspects may place the deliberations on an equal playing field (see also Sect. 4.6).

The conflict between USA and Europe, as it relates to GMOs, is seen predominantly on the international market. This is because North America is primarily involved in the export of GM crops, while the EU is an importer. The EU GMO regulations (see Sects. 3.3 and 3.5), implicitly based on the application of the precautionary principle, will ultimately affect trading agreements and costs of production. The unhappiness of the GMO-exporting countries was first publicly seen when the EU imposed a moratorium on GMOs in the late 1990s. When, in addition, traceability and labelling regulations were introduced, the USA filed a restriction of trade lawsuit with the World Trade Organisation (WTO). Such trade-related problems have the potential of developing into a real trade conflict.

A possible way to diffuse conflicting situations is by creating an agreed-upon international guideline for the use of GMOs. Two instruments are currently available: Codex Alimentarius discussions about GMOs and the already concluded Cartagena Protocol. Both are a step in the right direction to deal with GMO issues in a systematic and transparent manner. Codex Alimentarius is in the process of finalising new (risk related) definitions and standards for GMOs. As the standards agreed upon by the Codex Alimentarius are usually referred to by the WTO, this may open a way to reach binding international agreements (see also Sect. 4.1). In such a scenario, the basic ideas of the precautionary principle (and precautionary approach) could become part of the agreed upon Codex standards. This process is, however, time consuming with final guidelines expected in 2005. The UNEP Cartagena Protocol on Biosafety to the Convention on Biological Diversity emphasises the need to accept human and environmental concerns first and to accept that they take precedence over economic considerations.

4.2.8
General Discussion

The recognition that each country can take appropriate measures to protect human and animal health or the environment is generally universally accepted and is written into a number of international agreements. The precautionary principle includes these ideas as well. However, that does not make the principle the only correct way to deal with safety issues. Although most, if not all, nations may agree to the idea behind the precautionary principle, many will disagree on how and when to apply it.

The use of the precautionary principle on an international level faces four main challenges that have to do with drawing up appropriate terms of reference.

The first challenge is to clarify the role of science in the political decision making process, since political decisions are inherently not science-based. The other three challenges relate primarily to risk assessment guidelines and definitions that should increase our understanding of the scope and likely impact of the harm and thus help in the decision making process, namely:

- Defining the framework of the acceptable level of health and environmental protection against harm
- Asking the right questions and defining standards and methodologies to assess risks
- Defining an acceptable level of uncertainty and knowledge when providing answers to given questions

4.2.8.1
Precaution and its Relationship to Science

Supporters of the precautionary principle claim that it already contains a set of guidelines that define the framework of when and how to invoke and implement the principle. However, as already indicated in Sect. 4.2.6, it is important to understand the basis and intention of the use of precaution in the decision-making processes. Unfortunately, the word precaution has been misused and overused so that today we can see it capitalized as the Precautionary Principle or the Precautionary Approach. Precaution (the use of prudent action to obtain positive results) can be used in totally different situations with totally different aims and connotations.

When a scientist uses precaution, he/she may, for example, aim to come up with results and interpretations that will stand up to peer scrutiny, or aim to make sure that all appropriate safety and bioethical measures are in place. When precaution is used by a decision maker, he/she thinks within the framework of political decisions. They are driven by health, environmental and socio-economic concerns. The precautionary decisions of a scientist are based on science, while the decision maker takes science only into consideration. This distinction is very important. Decision makers make political, not scientific decisions. They do take scientific facts into considerations, but do not base their decisions on these facts alone. Indeed, the reason to apply precaution (as advocated by the precautionary principle) is because there is lack of coherent scientific data.

Work remains to be done to clarify the objectives and implications of mixing of science and socio-economic considerations in the decision-making process. Limiting the frames of reference of the precautionary principle to bio ethical considerations, may help to resolve this problem (see also Sect. 4.6). It is important for scientists to become actively involved in this deliberation process. One of the great challenges of the scientific community in the 21st century will be to define their role and identity within the society that is separate from other stakeholders. The danger faced by science and scientists is to be absorbed into, and made indistinguishable from, industrial or political agendas.

4.2.8.2
Defining the Framework of Acceptable Level of Health and Environmental Protection against Harm

Decisions on what constitutes an acceptable risk and tolerable level of uncertainty are very much country- and region-specific and dependent on local socio-economic conditions, moral values and collective experiences. The precautionary principle takes this into consideration. However, countries may disagree about the standards, let alone the concerns, values and experiences that should go into decisions about whether and how to apply the precautionary principle. In its present form, the precautionary principle does not attempt to reconcile these crucial differences. This seriously limits its international appeal.

Indeed, by incorporating socio-economic considerations into the precautionary principle and by placing the principle into the political realm, vested national or regional interests may be allowed to define harm, risk and uncertainty in a way that leads a priori to incompatible positions of what to do about a perceived hazard. The justification for these positions, according to the precautionary principle, is that they need to be in line with existing national regulations. As the example of precautionary principle and precautionary approach shows (see Sect. 4.2.5) this is not too difficult to do. The USA and Europe have different ways of dealing with risks and public protection. Both the precautionary principle and precautionary approach appear to follow the general principles of risk management that includes proportionality, consistency and non-discrimination. Yet they may be at conflict with each other, because of the key differences as outlined in Sect. 4.2.6.

These different positions need to be recognised and respected. Only then can consensus be reached about science-based acceptable levels of health and environmental protection and unacceptable levels of uncertainty as the basis of triggering the precautionary principle. What should be addressed are, in effect, definitions of "triggers of action" discussed by all the key stakeholders. Specific attention should be paid to harms that may be widespread, long-term, irreversible and accumulative. Input from scientists in drawing up the triggers of action should be based on a combination of state of current knowledge, risk and benefit comparisons to other comparable existing technologies, products or activities in cohort with a set of ethical criteria (see also Sect. 4.2.9 and Sect. 4.6). When agreed upon, this can be used to invoke the precautionary principle with a clear procedural check list to be followed. Indeed, this may become part of the normal risk assessment procedure as the case-by-case procedures evolve (see Fig. 4.1 and Table 4.2). Addressing these challenges is not necessarily the role of the precautionary principle. Yet they need to be resolved, likely though the Codex Alimentarius, to modify the principle for international use. As already eluded to, the first such attempt is likely to come from the Codex Alimentarius Intergovernmental Ad Hoc Task Force on Foods derived from Biotechnology. One of its main tasks is to set principles for risk analysis of foods derived from biotechnology.

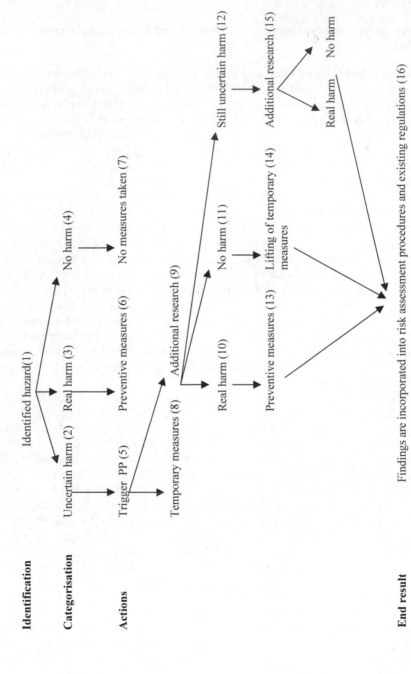

Fig. 4.1. Possible decision-making process using the precautionary principle (also refer to Table 4.2)

Table 4.2. Possible decision-making process using the precautionary principle as outlined in Fig. 4.1

1	In order to invoke the precautionary principle, the identified hazard (source of danger) needs to relate to possible risks to human health and/or the environment. New hazards are usually identified based on new scientific results (discovery of the ozone hole is a good example). The newly identified hazard can relate to emerging or already existing technologies, products or processes that may have already undergone risk assessment
2	Only in the case where there is uncertainty about the likelihood or impact of the risk associated with the hazard can the precautionary principle be invoked. Consensus needs to be reached on "triggers of action" to invoke the precautionary principle. The decision should be also preceded by a wide ranging discussion by all the stakeholders (see also points 8 and 9)
3	There is a real and definite risk to our health and/or the environment, immediate action needs to be taken.
4	Risk posed by the hazard is acceptable, no action needs to be taken
5	Precautionary principle can be invoked only on the basis of uncertainty about scientific results. Uncertainty can have a number of causes: divergent scientific opinions (for example about cause–effect relationships), insufficient data, wrong data, insufficient frame of reference (the type of questions asked and answered) etc.
6	Preventive measures are usually entrenched in existing regulations. Application of these regulations has to comply with national and international laws and principles (see Table 4.1)
7	If, based on existing information, the conclusion is that acceptable risks are associated with the hazard, no further action is needed
8	One of the two primary objectives of precautionary actions is to take temporary preventive measures to protect the health of the public and safety of the environment. There are a wide range of temporary measures that can be imposed, from outright banning to advisory statements. The choice of the temporary measures is made by the decision makers within the framework of national and international laws and principles (see Table 4.1). Although a political decision, it is advisable to take only bio ethical, rather than socio-economic, concerns into consideration when deciding on preventive measures, especially if used internationally
9	The second primary objective of precautionary actions is to carry out additional research to fill in the knowledge gaps. Its ultimate objective is to reduce the uncertainty about a possible risk to a level that allows a yes/no decision on whether and how to act (see points 3 and 4). The greatest challenge is for all the stakeholders to agree on frames of reference (the questions to be asked and answered), the tools and methodologies for analysis and interpretation of the results. Bioethical (rather than socio-economic) considerations should become part of the research design and considerations. The fear of many is that either the frame of reference will be too broadly or too narrowly defined, a priori open to subjective interpretations, rely too heavily on socio-economic political considerations after research is concluded and may protract the time needed for reaching a definitive yes/no decision. This may stifle research, freeze innovation and reduce rather than improve transparency and effectiveness of the decision making process. Additional research can have again three outcomes as defined in points 2, 3 and 4
10	The hazard poses an unacceptable risk to human health and the environment
11	The hazard poses acceptable risk and is no threat
12	Insufficient data is available to make a yes/no conclusion

Table 4.2 (continued)

13	In case of a real risk, a number of preventive measures can be undertaken, that are in line with national and international laws and principles (see also 6)
14	In case that the hazard does not pose a risk to human and/or environmental safety, the temporary measures imposed by triggering of the precautionary principle are lifted and/or modified
15	In cases where no definite conclusions can be drawn, additional research needs to be carried out and temporary preventive measures likely remain in place
16	The end result of invoking the precautionary principle should be to incorporate the lessons learned into the mainstream risk assessment and management procedures. This will in effect add additional filters and considerations to the existing decision making procedures for identification of new potential hazards (see point 1)

4.2.8.3
Asking the Right Questions and Defining Standards and Methodologies to Assess Risks

Once the precautionary principle has been invoked, the next challenge is to define questions that need to be asked to address possible risks posed by an identified hazard. The questions need to address realistic concerns related to health and/or the environment. The factors to consider and the questions to be asked will need to be formulated taking into account our tolerance to a certain level of uncertainty (see also next section).

For each instance when the precautionary principle is invoked, guidelines need to be decided upon to define what level of uncertainty is acceptable and what scientific evidence is needed to be reasonably certain that the identified hazard does not pose harm to the population or the environment. This is important, as the design, implementation and evaluation of each experiment normally bears the mark of individual scientists. As such, it is important to make sure that the questions, standards and methodologies to be used in risk assessment are well defined, reproducible, transparent and contribute to shedding more light on the risks in question.

This task is not easy. Baseline data to be used as controls (for example current agricultural practices) are often missing. This makes it difficult for a risk assessment specialist or manager to decide on the risk assessment methodologies to ascertain the differences, and thus implied safety, of the new product. As illustrated by the climate change debate, sometimes only correlative information may be available. This will preclude assigning reliable risk levels to a certain technology, product, methodology or activity.

It is clear that science and knowledge is evolving. The standards that we employ can only reflect our current scientific knowledge. In the case of GMOs, there is a need for risk assessment based on direct field evidence and for addressing complex environmental issues in an integrated manner whereever possible. With this approach, possible escape of genes from a GM crop may, for example, be viewed differently if the latest research can show the frequency of gene transfer between specific crops and wild relatives occurs naturally but that the likelihood of a new gene to survive in the new genetic pool is minimal. Conversely, the introduction of a foreign gene through genetic engineering may still affect the plant metabolism in ways we currently do not understand or predict. Scientific knowledge ultimately allows decision makers to take a definite decision on whether/how to allow a product onto the market.

4.2.8.4
Defining an Acceptable Level of Uncertainty in the Answers to the Given Questions

An ideal situation is when cause-effect relationships can be established without the need for the precautionary principle. However, when many interconnected questions are being addressed or questions are not well defined, it is likely that only correlative answers can be given. That means that it will not be sure whether A affects B, or B affects A or if something else affects both A and B. Correlations are not a good tool for risk managers to make final yes/no decisions. However, correlations can be a good basis for invoking the precautionary principle to generate more concrete cause-effect data. Once the precautionary principle is invoked, the questions to be answered should be defined in such a way as to give the decision makers the opportunity to reduce uncertainty to an acceptable level and make yes/no decisions. The questions should therefore aim to establish cause-effect relationships. However, the establishment of cause-effect relationships still does does not exclude the possibility of uncertainty.

Balance needs to be established between the need to expand our knowledge base and the realities of every day life. This means narrowing down the questions to such a degree to allow production of manageable chunks of data that can be analysed. It may not give answers to the big overall picture, rather it will provide clear answers to a subset of important, but limited questions. This would allow scientists to test results using the null hypothesis approach, in effect asking whether a given set of observations differs significantly from baseline observations. From these partial answers, an overall picture will eventually emerge, rather like putting pieces of a puzzle together.

Finally, the point at which it is possible to say with reasonable certainty that we have enough information is also contested. Even with sufficient knowledge and reduced uncertainty, conclusions about potential effects of GMOs and decisions about their use will involve an element of informed judgement. All answers will carry an element of uncertainty, so the question needs to be asked and partly pre-defined concerning how much uncertainty we can accept in order to say that risk has been reduced to an acceptable level. The challenge is to arrive at a flexible definition for acceptable level of tolerance for a possible risk. This will not be easy, as personal and societal experiences, needs and values come into play.

4.2.9
Conclusion

There are three outcomes in making decisions: yes, no and being uncertain. The precautionary principle is being applied only in the case of uncertainty. The precautionary principle allows risk managers to address early health or environmental warnings, weigh benefits and risks of action, and take an appropriate temporary precautionary response to minimise possible health and environmental damage. The precautionary principle expects decision makers to be on the side of caution. It is up to the country/region that initiates the precautionary action to argue why the action is necessary. The likelihood of imposing the precautionary principle should trigger discussions at all levels of the society to discuss specific and broader issues of the usefulness of, for example, the technology under scrutiny, alternatives available, questions of sustainable development and moral and legal implications.

There are two implications of applying the precautionary principle to situations of uncertainty. First, precautionary actions must be limited in duration because more new data is necessary to allow an objective final decision. Second, precautionary actions are biased to caution and so are not objective.

The precautionary principle should be invoked only in instances where concerns can be identified on the basis of science and ethics. Concerns need to be prioritised and must relate first to health and environmental risks. The challenge is to come to an agreement of what constitutes a hazard, harm, risk and threat as these are components of the triggers for precautionary actions. Although science is not always objective and results are subject to interpretations, science is nevertheless one of the best tools we have for reaching fair and transparent decisions that are respected across cultures. The primary aim of invoking the precautionary principle should thus be to reduce uncertainty in scientific results. Risk management decisions should be apolitical, carried out by independent risk managers and devoid of socio-economic considerations, apart from those embedded in ethical considerations of the specific case. Ethical considerations have a universal applicability and do address the questions of relationships (between people, people–nature etc) on a case-by-case basis. This ensures transparency, fairness and impartiality in the decision making process on an international level.

The terms of reference for precautionary actions should be defined according to the present status quo. Possible risks and benefits of a new technology or product should be compared with those of available alternatives, that in turn should also be subject to risk/benefit analysis. To concentrate on one technology, such as GMOs, without asking safety questions of conventional and organic farming is not correct. The relative benefits and risks of each technology should be compared as part of the risk assessment procedure. Caution should be exercised in how broadly terms of reference are defined. Science is unlikely to satisfy too broadly defined requirements. This may lead to indefinite postponement of introducing new technologies and innovations onto the market.

There is a need for urgent international action on a number of issues put into focus by the precautionary principle debate. Among them are agreements about various risk-related definitions (triggers of action), risk assessment procedures, standards, competencies, possible misuse of the principle and its oversight. Only when a genuine dialogue about the ideas and implications entailed in the precautionary principle are carried out with all the stakeholders on an international scene (perhaps emphasising science and ethics as its core) is there a hope of having the principle universally accepted and implemented. If universally accepted, the modified precautionary principle may even be applied more broadly and could, for example, be used in political decisions about managing world peace and international trade.

Finally, a few words about the impact of the precautionary principle on science. Although increasing our baseline knowledge, placing emphasis on risk assessment research may drain funding from general basic research that in a long run may prove more beneficial to our society. Overuse of the precautionary principle may lead to reduced emphasis on basic research and science in general. This may lead to science becoming reactive rather than proactive and creative. There is a possibility of harm to science, innovation and thus to the well being of our society. Thus applying the argumentation of the precautionary principle, the precautionary principle should be imposed to periodically evaluate its own impact on society and the environment.

4.2.10
Information Sources

Douma WT (2003) The precautionary principle: its application in international, European and Dutch law. TMC Asser Institute, The Netherlands

CCGP-Codex Committee on General Principles (2000) EU positions paper for Codex Alimentarius. The position paper can be downloaded from http://www.Eur-lex/en/com/index.html

European Commission's Communication on the Precautionary Principle COM (2000) 1. The communication can be downloaded from http://www.Eur-lex/en/com/index.html

European Environment Agency (2001) Late lessons from early warnings: the precautionary principle 1896–2000. Environmental issue report No.22

Foster KN et al. (2000) Science and the precautionary principle. Science 288:979–981

Kriebel D et al. (2001) The precautionary principle in environmental science. Environ Health Persp 109:871–876

Tait J (2001) More Faust than Frankenstein: the European debate about the precautionary principle and risk refutation of genetically modified crops. J Risk Res 4:175–189

von Schomberg R (2000) Agricultural biotechnology in the trade–environment interface: counterbalancing adverse effects of globalization. In: Barben D (ed) Biotechnologie-Globalisierung-Demokratie. Edition Sigma, Berlin

4.3
Public Perception of GM Crops

Paul Pechan, Giorgos Sakellaris, Anne-Katrin Bock

Several public surveys show variations in the perception of modern biotechnology. Medical applications are generally well accepted whereas applications in the agricultural and food sectors are viewed rather critically. This is true especially in Europe and there are several reasons for this, for example, health and environmental concerns, too little trust in governmental control regarding issues concerning GMOs, as well as the apparent absence of any advantage of genetically modified plants to the public at large. Additionally, most people do not feel particularly well informed regarding biotechnological issues.

4.3.1
Past and Present Public Attitudes Towards GM Crops

The 2003 Eurobarometer survey of the European public reveals widespread opposition to GM foods in much of Europe, but fears that the GM food debacle
would adversely affect public attitudes to medical and pharmaceutical applications of biotechnology prove to be unfounded. Europeans are not technophobic.
However, concerning biotechnology, they continue to distinguish between different types of applications, particularly medical in contrast to agro-food applications. The majority of Europeans do not support GM foods. These are judged
not to be useful and to be risky for society. For GM crops, support is lukewarm;
while they are judged to be moderately useful they are seen as almost as risky
as GM foods. By contrast, and despite the opposition to GM foods, perceptions
of medical biotechnologies (genetic testing, and the production of pharmaceuticals) and environmental biotechnologies (bio-remediation) are very positive.
Respondents were asked whether they thought specific applications of biotechnology were useful for society, risky for society, morally acceptable and whether
they should be encouraged (Table 4.3). According to their responses, they were
distinguished in three categories: supporters, risk tolerant supporters and opponents.

There are similarities and differences between supporters and opponents.
For example, around 80% of both groups say that they are "insufficiently informed about biotechnology" and similar percentages say that they would "take
the time to read or watch something about biotechnology in the media". The fact
that supporters and opponents alike feel poorly informed points to the need for
information campaigns, but equally that the effects of such information may be
uncertain.

Over the years, a number of surveys have been carried out to measure public
perception of biotechnology, including GM crops. During the height of publicity
campaigns against GMOs (1996 to 1999) there was a general increase in levels of
opposition to GM food in Europe. In industrialised countries, fear of the unknown,
concerns for the environment and food safety, moral or social considerations
and the question of freedom of choice were cited as influencing the attitudes
of the public.

Usefulness, the "Achilles heel" of the first generation of GM food products, is
a pre-condition for support. Indeed the absence of consumer benefits from GM
foods may amplify perceived risks and moral concerns. By contrast, where people
perceive biotechnologies to have substantial benefits, for example in health care,

Table 4.3. Three common logics

Logic	Useful	Risky	Morally acceptable	Encouraged
Supporters	Yes	No	Yes	Yes
Risk-tolerant supporters	Yes	Yes	Yes	Yes
Opponents	No	Yes	No	No

Source: Eurobarometer 58:0, 2003.

they are willing to tolerate risks (GM medicines and cloning human cells[1]). However, where biotechnologies are perceived to have only modest benefits, which come with modest levels of risk, there is no positive support (GM crops). In general where the public perceive no real benefits they are less willing to accept the perceived risks of new biotechnologies. Such technologies also suffer from a decline in moral acceptability and support.

Greater support for the cloning of human cells and tissues[1] than for the cloning of animals suggests that moral concerns are attached specifically to particular applications and not necessarily to underlying molecular biological techniques. Furthermore, greater opposition to GM foods than to GM crops suggests that, for the public, food safety outweighs environmental concerns. In the 2003 Eurobarometer survey, respondents were asked if they would buy or consume GM foods if they contained less pesticide residues, were more environmentally friendly, tasted better, contained less fat, were cheaper, or were offered in a restaurant. For all reasons offered there are more Europeans saying they would not buy or eat GM foods than those saying they would. The most persuasive reason for buying GM foods is the health benefit of lower pesticide residues, closely followed by an environmental benefit. Somewhat surprisingly, of the range of benefits included in this question set, price is apparently the least incentive for buying GM foods. Compared to other technologies, Europeans are less optimistic about biotechnology and genetic engineering, with the exception of nanotechnology and nuclear power (Fig. 4.2).

Related to the risk/benefit concerns is the question of choice. The Eurobarometer results from the 2001 survey on "Europeans, science and technology" showed that 94.5% of Europeans want to have the right to choose when it comes to GM foodstuff. The demand for choice is also echoed in USA, where the majority of people would prefer labelling of GM food products (Gallup and Rutgers University food policy Institute polls in 1999 or the ABC News polls from 2001 and 2003).

It is tempting to conclude that both the public in Europe and North America is misinformed or does not have the knowledge to make the "correct" decision about GMOs. After all, according to the 2001 Eurobarometer survey, only 60% of the population felt they understood the topic of GM food (compared to nearly 85% when asked about air pollution), and 85% of respondents wanted to know more about GMOs. Such findings have prompted calls to educate the public about such matters. Already in 1995, the American Dietetic Association stated that there was a need to educate the public not only about transgenic food, especially about food safety, but also about social, ethical, economic and environmental issues. This appeal to educate the public was repeated in many other reports and statements by the food industry, government departments, consumer associations and environmental associations.

However, the situation is more complex (see also Sect. 4.4). Thus although 85% of the public agrees that GMOs should only be introduced if proven scienti-

---

[1] Cloning of human cells or tissues to replace a patients diseased cells that are not functioning properly e.g. in Parkinson's disease or diabetes (Eurobarometer 48:0, 2003).

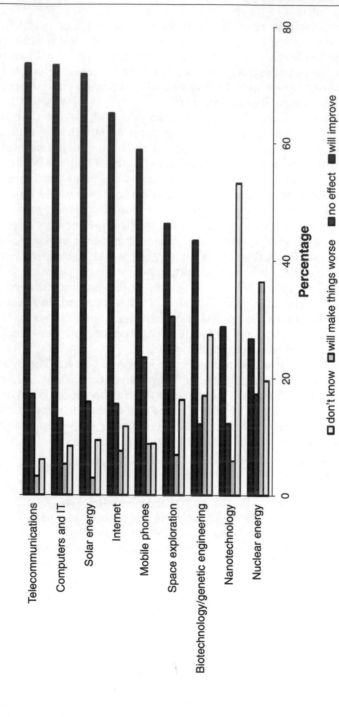

Fig. 4.2. Impact of technologies on the way of life. Source: Eurobarometer 58.0, 2003

fically harmless, this does not change the view of over 70% of the public that they would not want this type of food. Moreover, more than 60% believe that GMOs may have negative effects on the environment. The contradiction between accepting that GMOs could be introduced after undergoing appropriate scientifically based risk analysis on the one hand and the strong opposition to their use, suggests that:

1. The public is not knowledgeable or aware about the current stringent approval procedures for the release of GMOs onto the market
2. The public is concerned that the tests and procedures are inadequate
3. The public is expressing an opinion based on choice and concerns about the management of risk

The final report of the PABE project, addressing the issue of public views of GMOs, concluded that policy makers should be prepared to consider that the source of the problem is not only to be found in the behaviour (implying knowledge/awareness deficiency) of the public but also in the behaviour of institutions responsible for creating and managing innovations and risk. These findings confirmed thoughts already discussed for some time within the policy and decision-making circles in the EU. It is within this background that the decision was taken to label GMOs and insist on their traceability. The traceability decision is thus in part a confidence-building action in the transparency and decision-making process as it relates to safety issues (see also Sect. 3.3 and 4.1).

4.3.2
Awareness of the Issues Versus Understanding the Issues

Awareness of biotechnology centres mainly around the issues of GM crops. Although the awareness of GM food has increased, understanding of the technology has not followed the same trend. This implies that the public has reacted to the opinion-shaping activities of groups with vested interest in promoting their views, rather than becoming educated about the issues from impartial sources presenting both the pros and cons of the technology. In Japan, a survey conducted at the end of 1999 suggests that bad publicity concerning GM crops has even tainted the perception of other applications of biotechnology.

Although no comprehensive studies were published until the end of 2000, it is probably safe to say that, in Europe, environmental groups have been much more effective in getting their viewpoint across to the public than the industry itself. Interestingly, countries with media coverage and special interest group activities against GM food, have in general recorded higher negative public attitudes towards GM food. Partly in recognition of this the biotechnology industry invested over US $50 million in the USA and Canada, to embark on an advertising campaign extolling the benefits of biotechnology. This campaign is to be extended on a smaller scale in Europe, targetting primarily supermarket.

4.3.3
Issues in Public Discussion

There are two important aspects to consider when dealing with information on plant biotechnology: first, the content to be addressed, and second, the strategies of getting the information across to the public. The public will then be empowered to make knowledge-based decisions about plant biotechnology and GM crops, in particular.

4.3.3.1
Content

There are some core issues to be objectively addressed if there is going to be any hope of increasing public confidence in food safety and regulation of GM crops and food. The following topics are the centre of public debate on GM plants:

Environment	Uncontrolled spreading of transgenes, creation of super weeds, undesired effects on non-target organisms, impact on biodiversity and unknown long-term effects
Human health	Potential toxicity and allergenicity of foods derived from GM crops, spread of antibiotic resistance genes (used as marker genes) to micro-organisms within the human and animal gut, and the presence of GM ingredients in baby-food
Labelling	People want to know what they are buying and eating in order to have the freedom to choose
Access to information	Insight into the legal process for approval of GM crops, insufficient involvement in national decision-making processes and difficulty accessing relevant information as it is with the companies
Monopoly issue	Development of GM crops by a small number of very large multinational companies, dependency of farming on these companies, reduction of the number of crop varieties in use and protection of intellectual property through patents
Liability	Up to now the legal responsibility for any damages, potentially resulting from the use of GM crops, has not been solved
Ethical aspects	Breaking species barriers through gene transfer is perceived as unnatural, humans are "playing God", GM crops are superfluous because there is enough food and there seems to be no benefit in using GM crops

Additional points of contention relate to the way genetic modification of plants is viewed. For some, this is just another tool to be used by plant breeders. For others, genetic modification is seen as a fundamental change in the way that crops are produced. There is disagreement relating to the methodologies for risk assessment and the ability to trace GM material in the food chain. There is also little consensus on the method of marketing GM food.

If plant biotechnology, especially GM foods, is to be generally accepted, it must be clearly recognised by the public to have a benefit and be as safe as all other products available. Once recognised as such, perceived factual, ethical or moral objections could change or be dampened as has happened in the case of the application of modern biotechnology in medicine and the pharmaceutical industry.

4.3.3.2
Appropriate Strategies of Getting Information across to the Public

Short-term public awareness campaigns may bring some benefits, but as already eluded to, there is the danger of manipulating public opinion (persuading versus informing). A more appropriate approach is likely through long-term public education and dialogue programmes using reliable information sources and disseminators. It is, however, by no means certain, nor should it be expected, that this would increase the public acceptance of GM crops. For a more detailed discussion of communication with the public, please see Sects. 4.4 and 4.5.

4.3.4
Future Prospects

The necessity of involving the public in discussions on application of modern biotechnology and to take public concerns into account has been recognised at European level. In the White Paper on Food Safety, the European Commission expresses clearly the need for communication with the public, using public hearings and dialogues and also for interactive communication between the different stakeholders. Assumptions regarding the appropriate relations between science, society and government with regard to decision-making need to be reviewed. The EU aims to increase the efficiency and transparency of the decision-making processes and to involve the public in these processes.

The benefits of the first generation of GM crops are perceived as only serving the interests of the companies selling them and the farmers growing them. In contrast, the potential risks of these crops appear to be borne by the general public, e.g. by affecting the environment or the consumer's health. As such, the European consumers see little use in growing GM crops in Europe. Perhaps the willingness of public to accept certain risks in connection with GM crops will increase if it sees clear advantages in using this technology, for example, nutritional improvements, as is indeed planned for the next generation of GM food. This must be combined with a greater trust in the decision making authorities.

4.3.5
Information Sources

Eurobarometer 51.1 (1999) Europeans and the Environment. Conducted between April and May 1999, published in September 1999. Available at: http://europa.eu.int/comm/dg10/epo/eb.html

Eurobarometer 52.1 (2000) Europeans and Biotechnology. Conducted between November and December 1999, published in March 2000. Available at: http://europa.eu.int/comm/dg10/epo/eb.html

Eurobarometer 55.2 (2001) Europeans, science and technology. Available at http://europa.eu.int/comm/public_opinion/archives/eb/eb55/eb55_en.htm

Eurobarometer 58 (2003) Europeans and Biotechnology in 2002. Available at http://europa.eu.int/comm/public_opinion/archives/eb/eb58/eb58_en.htm

Kamaldeen S, Powell DA (2000) Public perceptions of biotechnology. Food Safety Technical Report no 17. The report is obtainable at: http://www.foodsafetynetwork.ca/ge.htm#public

Marris C (2001) PABE Report: Public views on GMOs: deconstructing the myths. EMBO reports 21:545–548

4.4
Communication Strategies in Biotechnology

Paul Pechan, Giorgos Sakellaris

Although not afraid of technologies in general, the public views plant biotechnology with scepticism. They see it as risky, not useful and morally questionable. Restoring the trust of the public will require not just better public relations strategies, but more profound changes in institutional culture and practice. They will need to demonstrate their capacity for adequate risk management through consistent behaviour. Information communicated with the public must be accurate, complete, easy to access, and understandable. Any communicator must use an appropriate communication channel in order to reach his audience; one of the most effective channels is the mass media, especially the broadcasting media. However, there is a basic conflict between the needs of scientists and science communicators on the one hand and the broadcasting media on the other. The need to get clear and powerful messages across to the public may lead to omitting important facts. Therefore, the way of defining for example a "successful" science-related television programme must be reexamined. Moreover, the conflicting objectives of the scientific and broadcasting communities, as far as the communication of science is concerned, need to be addressed.

4.4.1
Introduction

Debates on biotechnology issues have appeared in different countries and at different times across Europe: modern biotechnology has become increasingly sensitive socially and politically. In contemporary times, public opinion is not merely a perspective; it is a crucial constraint on the ability of governments and industries to exploit the new technology from the start. While the biotechnology industry assumed that regulatory processes were the sole hurdle prior to commercialisation, it is now apparent that a second hurdle, national and international public opinion, needs to be taken into account. In order to engage the public, a clear communication strategy must be defined. In such communication, science is viewed as a rational societal activity distinct from others, especially politics, but important for society if applied in a rational way to improve our living condi-

tions. However, not even scientists can foresee all of the benefit and risk implications of their work. This complicates the ability to clearly communicate science
related issues to the public.

4.4.2
Stakeholder's Perception

Most stakeholders in the GM debate misunderstand public responses to GMOs
and that this represents one of the key underlying causes for the current impasse
in the GM debate. The use of GMOs in agriculture and food has become one of
the most controversial topics in contemporary societies, especially in Europe.
Promoters of agricultural biotechnologies are concerned that the current public
controversy on GMOs in Europe is impeding the development and commercialisation of a new technological field considered to be of strategic economic importance for Europe. At the same time, critics who believe that GMOs involve
unacceptable impacts on the environment, health and society, continue to feel that
their concerns have not been fully addressed. The public is somehow caught in the
middle.

What do people in Europe think about the use of GMOs in agriculture and
food? What expectations and concerns do they have? How do they shape their
views when faced with a new issue such as this? How do they perceive this issue
within the whole context of modernisation and lifestyle changes? Characterisa-
tions of public responses to GMOs in decision-making circles are typically
framed either in terms of a lack of knowledge or of non-scientific ethical concerns, resulting in the appointment of expert ethical advisers or public consultations about the social acceptability of GMOs. However, policy makers need to
consider that the source of the problem is to be found in the behaviour of the institutions responsible for creating and managing innovations and risk, especially
as it relates to the issue of trust. Trust – or rather lack of trust – has increasingly
been identified as a key problem and issue to be addressed by policy makers
involved in risk management. The results suggest that trust is indeed an important dimension in public responses to proposed technologies and policies, but
that the way in which trust is most often conceptualised in policy circles is misleading and unproductive. Restoring public trust in regulatory institutions tends
to be seen as an issue to be resolved by improved communication strategies and
is largely treated independently from other policy decisions. The issue of trust
cuts across all the other socio-cultural factors identified. Restoring trust would
require not just better public relations strategies, but more profound changes in
institutional culture and practice. In order to restore trust, institutions would
need to demonstrate their capacity for adequate risk management through
consistent behaviour over a long period, and across different fields. This seems
the most urgent imperative for the development of a more constructive and satisfactory debate on agricultural biotechnologies in Europe.

4.4.3
Criteria of Science Communication Content and Choice of the Communicators

The form of the communication and the profile of the communicator are important considerations in determining the success of the communication activity. Information to be communicated must be accurate, complete, easy to access, understandable, not selective and precise. Also, several other communication criteria have to be taken in consideration. These are: social and ethical criteria, risk assessment and interpretation and impact of provided information.

The choice of the communicator depends very much on the degree of trust a potential communicator has with the public. Supporters and opponents generally agree that newspapers, ethics committees, doctors and consumer organisations are "doing a good job for society" (a proxy for trust) in respect to biotechnology communication. The public sees governments as aligned predominantly with industry in the promotion of biotechnology while the position of environmental groups tends to appeal to the opponents amongst the public. In contrast, newspapers, doctors and consumer organisations are seen as more impartial. A Eurobarometer survey demonstrated that around 70% of Europeans have confidence in doctors, university scientists, consumer organisations and patients' organisations. Around 55% have confidence in scientists working in industry, newspapers and magazines, environmental groups, shops, farmers and the European Commission. However, less than 50% have confidence in their own government and in industry. Despite the high trust, on the whole the scientific community lacks the skills to effectively communicate their science to the general public: in their professional role, scientists concentrate primarily on the production and dissemination of information to the scientific community. If scientists are to become involved in public debates on the purpose and implications of their scientific results, they must improve their ability to communicate with a wider audience.

Some of the issues affecting the role of scientists as communicators are:

- Truth
  The most important argument is that science reflects reality. We all tend to consider statements based on science to be the best approximations available. By implication, scientific experts often claim that their factual statements are truer than those from non-scientists. However, it happens that scientists often contradict each other altering the image of unity. They allege that their colleagues deliberately conceal risks or violate rules of equity, or disregard universally accepted ethical standards in the name of freedom of research. Such allegations often meet an open ear among an already suspicious public.
- Appeal
  Many scientists assume that not only they but everybody is (or ought to be) fascinated by science. However, we all know that science definitely does not appeal to everybody, and many people consider science to be incomprehensible, not least due to new terms constantly popping up. Scientific fields develop their own language and set of metaphors that people who have no particular interest in science do not understand, which can hardly be called a successful outcome of any communication strategy.

- Reputation
  Most scientists see themselves as hard working, responsible, well controlled and more critical than ordinary people. This self-image is hard to convey to the public in an unrestricted way and the reputation of the scientific community at large is put into question by cases like human cloning that violate moral issues.
- Lack of Ideology
  Scientific arguments are considered rational because, ideally, science does not take notice of interests, values, compromises and emotions. Science delivers best evidence as opposed to various ideologies, therefore, for rational decision making, scientific advice is indispensable. However, the argument of rationality is often suspected to serve as a proxy in order to exclude arguments and interests not shared by the majority of the scientific community.
- Common good orientation
  Scientists are convinced to work for the common good rather than for their personal profit. However, there are many profit-oriented interests involved in research these days. The scientists' common good orientation gets questioned not only in the light of industry interests but also from competition among scientists. Scientists are aware that this fact could jeopardise their communication strategy to defend their claims, so they often solve the problem by raising the argument that scientific findings will benefit patients and consumers, contribute to the country's economy and enhance competitiveness at large. However, the detour to the indispensability of science for a knowledge-based economy only make sense in the light of the international competitiveness paradigm.

4.4.4
Mass Media

One very important factor is that any communicator must use the most appropriate communication platforms in order to reach their audience. Mass media constitute a major forum of the public sphere in modern societies. There is general agreement in the literature that the mass media are influential, but less agreement about the nature of this influence. It is variously argued that the mass media serve to "frame" issues in the public domain, that they serve an "agenda-setting" role, and that they react to and therefore express public perceptions. In their role, the mass media not only serves as the communicator of issues but also provides the platform through which the communication is carried out. This double role sometimes leads to confusion about the tasks of mass media. As will be illustrated in Sect. 4.4.5, the core of the problem lies in unresolved conflicts between the different needs of the communication stakeholders and in the reliance of the mass media on ratings as a measure of their success.

Many scientists would claim that the media do not fully appreciate the inherent importance and interest of science, particularly basic science, and that journalists often have too little scientific education leading to less than optimal presentation of the information available. These views are perhaps not surprising. Scientists and news journalists, as members of two different social institutions, have different professional roles and information tasks. In the end, it is the media, not the scientists, who decide what is to be presented to the public.

4.4.5
Communicating Science Using Television as the Presentation Platform

4.4.5.1
Why Use Television?

In the 2001 Eurobarometer survey, nearly 66% of the population believed that there is not enough information on science and technology generally available. Only 60% of the population felt they understood the topic of GM food and 85% of respondents wanted to know more about GMOs. This, combined with the observation that nearly 90% of Europeans would like scientists to communicate their results better, clearly points to the need for the scientific community to improve communication of science matters in general, and GMO issues in particular.

How should scientists communicate with the public? The obvious candidate appears to be television, since 66% of people in the 2001 Eurobarometer survey indicated that this is their preferred means of obtaining scientific information. However, this choice creates a conflict of intentions because the objectives and reasons of scientists, to communicate science to the public, are not necessarily shared by the broadcasting media. Of primary concern to broadcasters are ratings. These influence which programmes are shown on television and how they are portrayed. Media people understand that bad news sells better than good news and brings in more viewers, higher ratings and consequently higher income from advertising. The challenge is to find a way to reconcile the need for genuine communication with the public on science matters with the broadcasters' needs for high ratings.

4.4.5.2
The Gap: Empowerment Versus a Good Story

As we have seen, there is a basic conflict between the needs of scientists and science communicators on the one hand and broadcasting media (television) on the other. This conflict can be considered in two ways: firstly in terms of the needs of both parties, and second, in terms of their professional constraints.

4.4.5.2.1
The needs of the scientific and broadcasting communities

The traditional need of the scientific community is to communicate experimental results, primarily to fellow scientists and granting organisations. This, in their view, assures them of professional standing and the ability to continue with their research. Unfortunately the dialogue between scientists and the public is often considered to be of secondary importance.

The broadcasting media need to present good stories to ensure that the public watch their television programmes. A good story usually means one that touches human emotions, one where people play a central role, and one that is stimulating to watch. Therefore, if there is a good scientific story that includes the above elements, broadcasters will, if they have space in their schedule, air the story on television. However, the need to maximise ratings means that, even if there is an important discovery, if it does not make a good story, it is unlikely that television will cover it in any real depth.

This gap between the need for public empowerment on scientific issues and a gripping story is sometimes impossible to bridge. Yet is it clear that the public needs to be informed and empowered on important scientific issues. Based on the 2001 Eurobarometer survey, nearly 80% of the public would like to know more about science while similar number states that scientists should do a better job in informing about the latest scientific results. As most of the public gets their information from television, it is clear that somehow the conflicting needs of science communicators and broadcasters need to be reconciled. However, in order to propose a solution, it is important to understand how television functions.

4.4.5.3
Professional Constraints: The Curse of Television Ratings

Television is business and making, or at least not losing, money is at the core of their concerns. The bottom line in television is how many people have watched a certain programme. This information is crucial because it helps television to argue with potential advertisers to get more money per minute of an advertising slot. Thus a programme that has 10 million viewers can charge more money per minute than a programme that has only one million viewers. In an extreme situation, some private television stations see television programmes as fillers for advertising. In other words, for many private television stations, content is secondary to high advertising revenues. But what about public television? As they are publicly funded shouldn't the ratings game and advertising play a lesser role? In theory the answer is yes, because public television, by definition is publicly funded and thus should not be dependent on ratings and advertising money. This should allow them carry out tasks that private television does not perform. In reality, public television is losing its identity because on the one hand they are publicly funded, but on the other hand also receive revenue from advertising, in direct competition with private broadcasters. In both cases, ratings play a major role. More viewers means more ammunition to argue with various granting agencies to continue their programming support and also increased advertising revenues. This emphasis on ratings influences the programme content. Only in countries that have an elaborate broadcasting system, such as in the UK, can the needs of the public be met though targeting of different groups with information or stories of differing levels of complexity, seriousness, and inspirational value. Many other countries do not have this structure and must make more stringent choices about what programmes to present. In the end, for both private and public television stations, science is a commodity that needs to be noticed in order to be placed in the programming schedules and adequately promoted by broadcasters.

4.4.5.4
Changing the Ratings Game

It can be argued that private television stations do not have the task to inform the public about science. Their task is to make money for the owners. However, public television stations are publicly owned and should thus have the ethical and moral responsibility to play a more proactive role in these efforts. They should make sure that timely scientific and technological information, for example, in the form of documentary films or debates, gets out to the public. This, however, means that the criteria of defining what is a successful programme must be re-examined. It is not sufficient to evaluate programmes only according to the ratings they receive. A more sophisticated approach is needed:

1. Information in a story needs to have a certain scientific content that is useful, complete, accurate, and retained by the public.
2. More money should be spent on developing science and technology program-mes, especially in support of script development since this is the foundation for accurate, complete, yet entertaining stories.
3. The choice of the target group should be better evaluated. Sometimes, certain issues can be explained better in schools. But television needs to be in the position to specifically target those no longer in schools: those in the work-force or retired.
4. There is an urgent need to add new rating evaluation criteria for science-based programmes. Current rating evaluations do not readily take into account criteria such as programme quality or educational (awareness) impact value. These criteria can make the evaluation of science programmes much more meaningful and consistent with the general need for proper science com-munication. Indeed, this additional information should make it easier to argue the case for more science programmes.

The above four points should be included into the overall programme evaluation. How much weight each of these points would carry in the final evaluation needs to be elaborated.

4.4.6
Mass Media and GMOs

Mass media often becomes the battleground to win the public opinion to support/ accept a specific aim or interest. GMOs are an excellent example of this struggle, illustrating the clash of conflicting aims and interests. On one side there is industry, represented by big international biotechnology companies, aiming to introduce GM plant varieties and food products onto the market. Opposing are some NGOs that want to stop GMOs as something unnatural, dangerous and unnecessary. Moreover, the media itself becomes involved in this contest and reacts to information published by other members of the media. As will be argued below, facts very often become the first victim of the media war.

4.4.6.1
Choices of Reaching the Public

In pursuing their agenda to oppose GMOs, larger NGOs can rely on an extensive membership where printed information in form of newsletters offers a very important mechanism to present their viewpoint to their audience and mobilise them into action. In order to reach and convince the broader public, some NGOs rely on the use of mass media (printed press, television or radio). Judging from public surveys the most effective means of reaching the public is through television. This knowledge has not been lost on NGOs; they have learned to effectively gain the spotlight in television news by carrying out very specific and visually grasping actions.

4.4.6.2
Mechanism of Getting a Visual Message across

The choice of message content presentation, in a way that the public not only notices but positively reacts to it, relies on the understanding of human psychology. Advertising experts and film makers have known for a long time a primary way to reach an audience is through their emotions. This allows the audience to become receptive to the message and to react in a desirable way.

The general strategy is to concentrate on specific messages in form of short high impact packages. In analysing the psychological basis of creating an effective visual message three aspects need to be considered. First, the message content, second the means used to bring the message across to the public most effectively and third, viewer's perception and reaction mechanisms.

4.4.6.2.1
Message Content

The message can be for example that "GMOs are artificial and dangerous for our society and the environment". This simple statement can represent the complete message (story) to be brought across to the viewer. Although an important statement, we may or may not react to it unless presented in a way that catches our imagination, our heart and our attention. This is where the people whose job it is to package the core message into appropriate emotional clothing "rollercoaster" come into play.

4.4.6.2.2
Dressing up the Message

If the message is brief (with no time to develop its own storyline) the choice of how the message is presented becomes very important. Often this is done intuitively by the individuals who have come up with the core message. However, it is public relations or marketing departments that are usually responsible for bringing the core message across most effectively. The aim is to focus the viewer's attention on the key message in a way that there is no escape from it; they simply must absorb it.

The message "GMOs are artificial and dangerous for our society and the environment" has taken on the clothes of Frankenstein so that GM crops became "Frankenfood". The image of Frankenstein embodied both the folly of human activity and the dangers of trying to bypass nature. Most of us are familiar with the story and the images of Frankenstein from movies. The choice of this image was brilliant as it created a lasting association in the viewer between GM food and its unacceptability. Thus the choice of words, the context used and how they are presented can have a significant effect on public attitudes to GM food.

4.4.6.2.3
Viewer's Reaction

Sensory stimulations constantly compete for the attention of our conscious mind. All of them are initially processed at a subconscious level. We become aware only of those stimulations that are deemed to be important. When faced with a limited time on television, it is of utmost importance to catch the attention of the viewer for as long as possible. This is achieved primarily with powerful images, symbols and musical stimulations of our senses and cognitive processes. Dialogue initially plays a secondary role.

The story of Frankenstein is transferred to the viewer with all of its implications. The message dressed up as Frankenfood is perceived initially by his or her subconscious. As the story of Frankenstein is generally recognised in the western world, the images are likely to get the attention of the viewer, who can then become consciously aware of the complete message. The viewer's conscious mind, utilising lifelong experiences and attitudes, will contemplate and ultimately colour the message in a way that results in a decision as to whether (or to what extent) to agree with the message. Often, the viewer forgets or is not aware that the message is presented in a way so as to trigger a specific response. Therefore, powerful stimulations will elicit feelings, wishes, experiences and attitudes, without necessarily entering our conscious mind. In the case of "Frankenstein food", there is the possibility of elicitation and transference of the viewer's childhood fears of the unknown, of darkness, of death. The story of Frankenstein represents and invokes memories and feelings of pity, fear and anger: fear and pity of Frankenstein and anger for the folly of his creator (represented by the industry). Similarly, the images of airplanes hitting the twin towers of New York have become an instantly recognisable symbol of fear of unpredictable violence and pain. Placed in an appropriate context, this can create a powerful elicitor and link to the message we wish to convey.

4.4.6.3
Do Means Justify the End?

In presenting a message in the most effective manner, it is easy to stray away from objectivity and truth. In the case of GM crops, the original concern for the safety of the environment has been transformed into a negative image of fear and caution. But, as other chapters in this book have shown, only some currently grown GM crops in certain regions may sometimes pose a problem to certain insects. Moreover, all GM food on the market today is safe to eat. Does this then justify labelling GM crops as Frankenfood?

4.4.6.4
Morality of Telling Partial Truth

The argument of some NGOs is that industry is very powerful and one needs to use whatever means to fight it, if convinced of their wrong-doing. It is of course true that industry has large financial resources for public relations campaigns but this is also true for large NGOs. Moreover, as the case of the Frankenfood message illustrates, it is easier and cheaper to negatively label a product than to laboriously win a public support for its use when its benefits to the consumer are not immediate or obvious. This is not to say that industry is innocent and completely honest with the public. Industry would have been happy to see the introduction of GMOs to proceed quietly and quickly so that they could have argued that it is too late to stop the process.

Another argument raised by some NGOs is that today it is not possible to tell the complete truth because television news requires stories or messages to be told in less than 20 sec. Within that short time span, one needs to drive the message home hard with the effect that it may not represent the entire truth and may be somewhat misleading. Indeed, a lot of NGO effort goes into making sure that the messages presented cannot be legally challenged.

Nevertheless, some NGOs would be wise to review their public relations policy. It is simply not moral nor ethical to make misleading statements, even if they feel that they fight for the right cause. They risk losing the respect and trust of the public.

4.4.6.5
The Real Struggle

GM crops represent many of the political issues some NGOs and political parties are fighting against. The fight for public opinion about GM crops has to do not only with the safety of GMOs but also with issues of free trade, the power of industry, globalisation, sustainability, fairness, monopolies and rights of individuals.

Some NGOs have chosen to remain outside of the political arena to be able to advocate specific causes that political parties may not find attractive enough to pursue. This is very commendable. However, when an NGO begins to follow a political agenda, as listed above, they sometimes cross a line that makes them indistinguishable from a political party and thus become accountable to all the voting population.

4.4.7
The Way Forward

There will always be the need for NGOs and consumer associations to be independent overseers of the our society and the champions of the weak and needy. Sometimes this may lead to conflicting situations. This is also healthy in a democratic society. Yet it may be that on important issues that may change the future of our planet, this mechanism may not be sufficient as means to ensure checks and balances within the public arena. The need to get clear and powerful messages across to the public may lead to omitting important facts in the debate. It therefore may be necessary for governments, together with the key stakeholders (especially the public, NGOs, media and scientists), to discuss key (emerging) issues proactively in a more structured fashion. First and foremost this means being able to convince the public of the urgency and importance of the chosen topics. Second, it requires that sufficient time and effort is dedicated to discussing these issues in a way that will sustain the public interest and engage them in a constructive dialogue. The dialogue should concentrate on engaging the public in activities to minimize unfounded concerns, maximize empowerment and increase trust in the dicision making processes. It should complement, but not replace, the duties and responsibilities of elected officials.

4.4.8
Information Sources

Belton PS, Belton T (eds) (2003) Food, science and society: exploring the gap between expert advice and individual behaviour. Springer, Berlin Heidelberg New York

Eurobarometer 58.0 (2002) Europeans and Biotechnology. Final Report to the EC–DC Research Project LSES

Fjaestad B (2001) Why journalists report science as they do. Proceedings Communication Strategies in Biotechnology, Athens–NHRF

Gaskell G et al. (2000) Biotechnology and the European public. Nature Biotechnol 18:935–938

Gaskell G, Bauer M, Durant J (eds) (2002) Biotechnology: the making of a global controversy. Cambridge University Press, Cambridge

Marks LA, Kalaitzandonakes N (2001) Mass media communications about agrobiotechnology. AgBio Forum 4:199–208

Marris C (2001) Public views on GMOs: deconstructing the myths. EMBO reports 21:545–548

Marris C, Wynne B, Simmons P, Weldon S (2000) Public perceptions of agricultural biotechnologies in Europe (Final report of the PABE–EC–FAIR). Commission of European Communities, Brussels

Science Museum, London (eds)(1999) Biotechnology in the public sphere. Science Museum, London

Science Museum, London (eds)(2001) Biotechnology: the years of controversy. Science Museum, London

Torgersen H (2001) Remarks on scientists' communication to the public. Proceedings Communication Strategies in Biotechnology, Athens–NHRF

4.5
Communicating Risk

Paul Pechan, Giorgos Sakellaris

Decision makers, at all levels, must always weigh the risks against the benefits of their decisions. Today, many of the technologies we use are complex. Increased knowledge of how these technologies work often raises uncertainties, especially when the technologies have multiple applications. Decisions, when based on inherent uncertainties, increase the level of possible risk. Precautionary actions may be called for as well as an in depth risk assessment in order to eliminate or reduce the uncertainty about the risk and its effects. Constructive dialogue between the key stakeholders is essential for becoming aware and familiar with the concepts of risk and uncertainty. It is also important to address specific technologies, such as genetic modification of crops, where real and perceived risks may influence our decisions. There are many platforms where this type of communication may take place, for example through the media or in schools.

4.5.1
The Challenge of Communicating Risk Issues

Today's world is very complex and our society is faced with an interplay of physical, environmental, economic and social risks. This is primarily because of the ever-increasing integration and globalisation of our society. Such risks may have multiple trans-boundary impacts over various extended time periods. The OECD refers to these risks as systemic risks. Systemic risks are at the crossroads between nature-based events, economic, social and technological developments, and policy-driven actions. The different dimensions, interaction and prevalence of such risks in our daily lives complicates the communication of risk issues to the public.

The public knows or intuitively feels this new evolving pattern. The way that the public may perceive and interpret a certain risk can have a great impact on what scientific projects are carried out and what scientific applications will eventually be brought to the market place. For example, various areas of biotechnology, such as GMOs, are seen as highly controversial, not only because of moral concerns, but also because of how risks are perceived, defined and dealt with in this rapidly developing field of science. Thus, the decisions in the sphere of research and its application, as they relate to risk assessment and management, are often questioned by a worried and concerned public. In addition, the risks of the same technology may be perceived differently depending on its application. Genetic engineering applied to human diseases is, with a few notable exceptions, generally welcomed by the public, yet the benefits of genetic engineering of plants is not as obvious to the public and has been generally opposed on grounds of potential environmental and health risks. Ethical and economic considerations, as well as the global character of these issues and the vested interests of the key stakeholders, complicate the communication and decision-making processes.

4.5.2
Why Communicate Risk Issues?

Provision of health and a safe, clean and sustainable environment to the people of Europe is one of the cornerstone objectives of the decision makers within the European Union. Decisions must be made about what is safe and beneficial for our society and the environment. However, any decision to improve our well being is linked with weighing the risks and benefits. At the decision-making level, these include the analysis of risks, uncertainties and possible precautionary actions.

The complexities of risk deliberations, the various definitions and risk dimensions to be considered are often underestimated. As risks always contain uncertainties, proper choices and decisions can be difficult to make. The level of uncertainty about the real or perceived risk can lead to a variety of attitudes and decisions depending on the perspectives taken. This makes communication between experts and the public extremely difficult. Risk can be defined as unforeseen negative effects or consequences of a potential hazard, and can be expressed as the likelihood of a negative event (harm) occurring, combined with the magnitude of impact it will have if it occurs. The effects of the event can have a direct or indirect impact with immediate or delayed action. Moreover, the complexity of dealing with risk issues is compounded by the intricacies of risk governance.

Risk governance is composed of risk assessment, risk management and risk communication. Risk assessment has a number of stages: identification of the risk, its evaluation (potential impact, likelihood of occurrence and potential risk value) and decision on the risk management strategies to be taken. Risk management is composed of many elements: confinement strategies, restricted use, monitoring, traceability and following approved procedures. These are obviously not easy concepts to explain, yet there is clearly a strong need for the process of communicating risk-related issues to take place. Added to this is the challenge of communicating complex scientific topics such as those related to GMOs.

Constructive dialogue with the key stakeholders is needed to minimise unfounded concerns and maximise public empowerment on these complex issues while, at the same time, allowing a positive feedback to assure that science is moving forward in cohort with social needs and expectations. These societal expectations are themselves based on personal attitudes and decisions of "what is it that I really want and where do I stand on this issue?" The deliberations include weighing the risks and benefits of taking or not taking an action or decision. Thus, the dialogue about risk issues stands at the very centre of a democratic society. The deliberations reach all of us, both as individuals and as a society.

4.5.3
Who Should Communicate Risk, and with Whom?

The decision of who should communicate risk issues depends on the target groups. Indeed, the choice of the target group will also dictate the content of what should be communicated and how it should be communicated. It is thus imperative to choose the correct target group as a lot of subsequent effort depends on this decision. All parties concerned should be in a position to learn from the debates. The likely stakeholders, either as communicators or as target audiences, can be divided into three broad categories. Those that have something to communicate, those that wish to know more and those that need to know more. It is very likely that stakeholders can belong to more than one category, depending on the topic and immediate needs.

The category of communicators of risks can contain scientists, industry, decision makers, NGOs, media, communication specialists as well as representatives of professional associations such as doctors. Within the category of wishing to know more are predominantly groups within the general public that could be subdivided for example on the basis of their age, sex, interests, education, income, language or attitudes. The category of those that need to know more includes decision makers and others who through their profession need to be informed about risk issues.

In case of GMOs, the main stakeholders are scientists, industry, NGOs, general public and decision makers. If the target group were the general public, the method of presentation and content would differ from that if the target group were school children. In the former case, it may be important to solicit the services of a high profile public figure to communicate the key messages. The likely platform would be television (see Sect. 4.4). In the latter case, teachers would be the communicators. Indeed, teachers would also be the target group as they would need to be brought up to date on key issues and topics. The communication platform would be the classroom, using books and videos as teaching aids.

Due to the importance of the topic, risk communicators must be well chosen. They must be in the position to communicate complicated issues and uncertainties in a clear and explicit manner and, as risk issues deal with decision making and emotions, to be sensitive to the concerns raised. The communicators must also be able to listen, learn and draw conclusions from interactions with the various target groups.

4.5.4
What Should be Communicated?

Risk analysis is a systematic way to more fully assess risks, to get transparency into complexity of issues and to address uncertainties or knowledge gaps. In a food context, risks involve potential impacts on consumers. While all effort is made to minimise harm occurring, food safety is not an absolute and hazards do exist. Risk assessment follows a structured approach to estimate the risk level of an identified hazard and to obtain insight into the factors that influence the risk in a positive or negative sense. Risk managers decide whether a risk assessment is needed and support the risk assessors in their work. When deciding upon the best way to manage a risk, and when implementing their decisions, communication between risk managers and the public and private sectors is very important. As the concepts are difficult to explain, it requires an expert risk communicator to tackle these issues in a public forum. The discussion does not need to be technical, and may for instance address economical, social, and ethical issues of public concern.

A key aspect in the dialogue with the public is to be aware of how risks are perceived by the audience. Risk perception refers to a wide array of primarily psychological studies, examining why people perceive some risks differently than others. People are more concerned about involuntary than voluntary risks and more about technological than natural hazards. The original risk communication strategies worked top-down, for instance from a regulator to the public. More recently, a dialogue form of risk communication which encourages public and stakeholders to actively participate in the communication process has become more popular. Risk perception is influenced by a variety of factors, such as the ethnic, social, or race origins of the audience.

In order to be effective in risk communication, three main risk communication objectives must be taken into consideration:

- Increasing awareness and understanding of issues
- Increasing risk tolerance in appropriate cases
- Considering a democratic context of communication

4.5.5
How to Communicate Risk Issues

Mass media, schools and professional organisations are the likely platforms for communicating risk issues. The choices do not differ greatly from those already described and discussed in Sects. 4.3 and 4.4. It should suffice to say that the choice of television can effectively address the general public. However, in order to engage in a dialogue and open the possibility of active involvement of the public in discussing risk issues, more elaborate public consultation infrastructures will need to be put into place.

4.5.6
Information Sources

Blaine K, Powell D (2001) Communication of food-related risks. AgBio Forum 4:179–185. The article can be downloaded from: http://www.agbioforum.org

EFOST (2001) Report: What is Risk Analysis? EFOST, Ankara

European Federation of Biotechnology (2003) Who should communicate with the public and how? Report of the focus workshops. European Federation of Biotechnology, Delft

OECD (2000) Emerging Systemic Risks: First Meeting of the Steering Group. Summary of the Discussion and Conclusion. Minutes 90377. OECD, Paris

Renn O et al. (2002) Systemic risks: report to the OECD. Center of Technology Assessment, Stuttgart

Thompson PB (2003) Value judgements and risk comparisons. The case of genetically engineered crops. Plant Physiol 132:10–16

4.6
The Morality and Ethics of GMOs

Paul Pechan

The arrival of biotechnology and specifically genetic engineering has opened up a wide range of possibilities for exploring and using this technology, for example in medicine, agriculture and environmental sciences. The application of genetic engineering to agriculture allows manipulation of plants and animals to an extent never seen before. It is now possible to insert well-studied genes into genomes of unrelated organisms with well-formulated objectives. Understandably, concerns have been raised about the safety, usefulness and morality of this technology.

Public concerns are expressed in a number of forms, from personal expressions of mistrust of the new technologies to public protest actions by environmental groups in support of alternative agricultural practices, such as organic farming. Many people object to genetic modification of plants because of their deeply held views based on religious upbringing, life philosophy or long held social values. They may claim that this technology is unnatural and that we are playing God. Others may object to the technology based on information they have received through media, family, friends or school. They may fear for the environment or safety of the food products. Ultimately, individuals and governments formulate their concerns as opinions that lead to decisions on the extent of acceptance of genetic engineering. The opinions and decisions that we formulate are an expression of our perception of reality, as influenced by the interplay of our conscious and unconscious self and based on life experiences and accumulated knowledge. Opinions and decisions help us to deal with specific concerns in a way that reflects the moral principles of the particular society, culture, race or ethnic group to which we belong. The decisions at a political level on how to address and resolve these concerns are complicated because they have potential political, economic and social implications.

This section explores how personal decisions about GMOs are arrived at. The aim is to raise the reader's awareness of the components that influence the personal decision-making process.

4.6.1
Use of Facts and Morality in the Decision-Making Process

Scientists are used to discussing the latest findings, uncertainties, possibilities and scenarios as part of their quest for knowledge and understanding. Scientific findings are peer reviewed and published in scientific journals. These published findings are the "facts" that are expected to be used as the basis of discussions and decisions, for example about GM crops. It would be expected that a general consensus would be found as to how best to proceed with GM crops in the future. Yet, there are often deep differences in opinions and conclusions drawn. These differences result from how the facts are interpreted and the plurality of moral attitudes not just within the scientific community, but within human society in general.

The two dimensions of a decision-making process are factual and moral considerations. It assumes awareness of one's actions. Factual information is defined as information that can be independently measured, tested and verified. In most cases scientists agree on the results. In some cases, the level of uncertainty about the results or competing theories allows different interpretation of the new information (see also Sect. 4.2). Moreover, a decision-making process based on new information provided by scientists does not exist in a vacuum. It functions within a certain moral framework.

Morality helps us to distinguish correct and incorrect choices and actions, and is influenced by (for instance) religion, philosophy, regional societal values and beliefs, ethnic background, culture, and race. Moral attitudes and concerns can influence whether, and to what extent, facts are accepted or rejected as incomplete or faulty. Two types of moral attitude are described below (see also Sect. 4.6.5).

4.6.1.1
Fundamental Moral Attitude

This view can, for example, state that GM crops are wrong in themselves – we should not manipulate nature and/or we should not be playing God. The stimulus to express these concerns comes from deep internally held views that are influenced, for example, by religion or philosophy value systems. Some religions and environmental movements are representative of these two views. Decisions based on internally held views, without taking heed of factual information, are usually highly subjective. They are difficult to discuss on factual grounds as very often this may threaten the perception of reality of those that hold these views. Decisions on purely internally held views are nevertheless a valid expression of an individual, as long as these views also respect the rights of other people to hold other views.

4.6.1.2
Moral Attitudes That Take New Information into Consideration

This view may be expressed, for example, in a statement that GM food is not acceptable because of possible unforeseen environmental consequences. It would thus be morally wrong to release GMOs into the environment before we have more facts about the impact of GMOs. The stimulus to express these concerns originates from our daily exposure to information and opinions and often relies on personal risk and benefit analysis. The problem encountered here is to know whether the person actually has access to sufficient factual information or whether the decisions are based on a combination of partial information and observations enacted through the moral framework, leading to what is often called a "gut feeling" response. Education, personal interests and social conditioning play a major in such decisions. This may lead to situations where one person may argue for GMO (such as a farmer growing GMOs) while another may predominantly see the risks (such as some environmental groups). Each filters the factual information to suit their predisposed needs or moral views. Our reactions may thus be a mix between the conscious and subconscious processes of working out and interpreting the problem at hand.

The important field of bioethics has arisen and been established primarily as a response to address and resolve what may sometimes be conflicts between factual information or considerations and morality. It examines both in terms of appropriateness of choices and actions. It is a subject where science, philosophy and law meet and deals with the conditions and constraints under which we should apply new biotechnologies. The principal issues dealt with by bioethics include:

1. Weighing out the risks and benefits
2. Placing restrictions on research and its application
3. Rights and needs of humans/living organisms

4.6.2
The Broader Picture

4.6.2.1
Responsibility and accountability for actions taken

A large proportion of the public in Europe objects to genetic engineering of plants. The public feels that this new technology is different from traditional plant breeding. The products must be viewed with greater caution than those originating from conventional or organic farming. Proponents of GMOs would argue that an effective negative campaign of plant biotechnology opponents has scared the public into rejecting GM food products. Critics of GMOs would argue that the opposition is normal in view of possible environmental and health risks (see also Sect. 4.1).

Genetic engineering of plants differs from the previous attempts and successes of plant breeding. Unlike traditional plant breeding, genetic engineering of plants can greatly speed up the time from discovery to application and also potentially speed up and permanently change the pattern of evolution to an extent never seen before. A key aspect of this technology is the precision with which genetic changes can be made and the outcome predicted. Because we can now change nature with increased precision, we also have a much greater responsibility than in the past to decide how best to use this technology. For each plant that is genetically modified, scientists know more than ever before what is to be expected from the new transgenic plant and how it should and could behave under field conditions. Although knowledge gaps do exist, we can no longer claim ignorance of the possible effects. That knowledge carries increased ethical responsibility for the consequences by all parties concerned. It was partly the disappointment because decision makers and industry did not seem to rise to this challenge of responsibility that led to protest actions by some environmental and consumer groups and brought the genetic engineering debate to the attention of the general public. In the absence of direct benefits to the consumers, the debate centres around the moral and ethical responsibility of protecting human health, our social values and preventing abuse of the environment.

The level of responsibility and accountability is based on the amount of knowledge about the risks a hazardous event may pose: (a) how severe the impact will be if the event does occur and (b) the likelihood of the event occurring. The amount of knowledge is based predominantly on risk analysis of the given hazard. The more predictable and severe the outcome, the more moral and ethical responsibility decision makers need to carry. Thus, if the hazard were identified as posing catastrophic consequences to the environment with a high likelihood of occurrence, decision makers would need to take preventive action. On the other hand, if a very unlikely event with low impact does occur, decision makers could be excused for not having taken preventive actions. An interesting decision-making scenario is what to do if the probability of a hazardous event is small but impact would be significant, or when uncertainty remains about accumulated scientific evidence. Here choices based on moral and ethical considerations do become very important in weighing social, economic and environmental needs and priorities. In Europe, this situation may invoke the precautionary principle, a decision-making tool that can be invoked whenever facts are considered insufficient to take a conclusive action (see also Sect. 4.2). Depending on the circumstances, this may help decision makers to either initiate, postpone, or block an action.

4.6.2.2
Are there Universal Standards of Moral and Ethical Behaviour?

There are some aspects of moral attitudes and behaviour that are universally accepted as standards, others may vary region to region. A good example of a universally accepted standard is the UN declaration on human rights. Such standards can be traced back to our roots in religion, philosophy and what could be termed universal awareness of respect for other living organisms. Yet it is clear that what may be morally acceptable or encouraged in one part of the world, may be viewed elsewhere as morally questionable. One such example is the use of DDT. It is not allowed in Europe because it can damage human health and the environment. Yet the use is allowed in some African countries to control outbreaks of malaria-carrying mosquitoes, reducing human fatalities. Morally it may be argued that the use of DDT is wrong because it harms the environment. Yet both morally and especially ethically it can be argued that DDT use may be excused in specific cases because it saves lives, especially if no effective alternatives are readily available. Therefore, all decisions need to be carried out within the moral and ethical dimensions that must always be placed in the right context on a case-by-case basis.

4.6.2.3
Why Application of Genetic Engineering in Medicine Does not Meet the Same Resistance as its Application in Agriculture

It has been argued, especially by industry, that acceptability of GM crops will increase when real benefits to the consumers become apparent by introducing added value GM foods that have, for example, increased vitamin content. However, if the decision makers are still viewed by the European public as not responding to the challenge of carrying greater responsibility, accountability and transparency (all components of the moral and ethical dimension), the public will continue to have little trust in the decision-making institutions and GM crops. This attitude of lack of trust has been shown again and again by various public opinion surveys. Thus, as long as the public feels that decision makers do not act on a moral and ethical basis, and therefore cannot be trusted, it is doubtful whether GM-derived crops will be ever be extensively grown in Europe.

The situation is different in the application of genetic engineering in medicine. Here, benefits to the patients are clearly defined and the medical ethical guidelines well established. Because medicine has been practised for thousands of years, there is a well developed system of checks and balances to avoid misuse and mistakes. In medicine, the key relationship is between the doctor and the patient. The doctor is obliged to work for the well being of the patient, with respect and fairness. Thus, a doctor who chooses to damage the patient's long term health can be brought to justice. Doctors, rather than governmental officials, are seen as the decision makers with personal responsibilities. Of course, advances in human genomics bring with them new challenges that have a broader societal impact and where governmental regulations and involvement is likely. The controversial question of whether we should allow cloning of people has met with an overwhelming rejection by our society on moral grounds that incorporate ethical and scientific considerations. This has led to laws regulating the use of cloning.

Plant biotechnology is not like medicine and the ethical rules applicable for medical profession, although a good starting point, are not satisfactory for plant biotechnology research and its application. There are too many variables and possible interactions between the environment, the consumer and the producer. Unlike medicine, plant biotechnology is currently driven by economic considerations, not the health of a patient. Moreover, there are no universal rules of business conduct yet agreed to and no enforcement exists.

4.6.3
Deciding Whether GM Products are Safe and Useful for the Consumer

Philosophically, there are two opposing views when dealing with GMO safety issues. The first position states that if we forbid something we should have a very good reason to do so. This means that the burden of proof would be with those that wish to prevent the release of GM products. The second position states that we should not allow something to be released, unless we can show that it is safe. In the European decision-making process, proof of GM product and food safety lies with the applicant. They must convince the decision makers and the public that the product is safe before it is placed onto the market. They do this primarily on the basis of a comparison with other foods or agricultural practices. However, opponents of GMOs claim that this proof only shows lack of harm and does not actively show that the crops or products are safe. If one accepted this argument then everything that we eat or use today would need to be re-analysed. Our society would come to a standstill. For practical purposes, the approach adopted for the first generation of GM products is based primarily on extensive toxicity and allergenicity tests, combined with an investigation into whether the GM products are more or less harmful than similar non-GM products or practices on the market (see also Sects. 3.5 and 3.6).

The following two examples illustrate how different decisions can be reached on the safety and usefulness of GMOs, based on different interpretation of facts and by emphasising different moral and ethical values. The terms of reference for the following examples will be the impact of GM products on the health of consumers.

4.6.3.1
Scenario A: Arguments for GM Products

4.6.3.2
Fundamental Moral Attitudes

We have already been manipulating nature for thousands of years, including imposing profound genetic changes on species. Only a few things we eat today have not been touched by human selection. If natural means "as God has given to us" or was found in nature before the onset of farming, then nothing we eat today is natural. It would be unfair to say that now, with the onset of biotechnology we are suddenly changing nature and playing God. We have always changed nature for the benefit of human kind. There is no problem with crossing the species barriers because at a molecular level, a gene is a gene. Genes are composed from the same building blocks. Transfer of such genes represents nothing more than using available natural components for the benefit of human kind. In addition, horizontal gene transfer between plant species can occur, albeit at low frequencies. What counts is to analyse the effect of these genes after they have been inserted into a new plant. The above attitude is formed primarily by trust in science and evolution.

4.6.3.2.1
Moral Attitudes that Take New Information into Consideration

No amount of testing will show that GM products are 100% safe. The safety of GM products can only be shown relative to other products or processes. Indeed, no food we eat, GM or not GM, can be labelled as 100% safe. GM foods on the market today are by far the best studied and examined of any food we eat. The quality and fairness of decisions depends on the methodology and the criteria used in risk assessment. Methodology determines the quality of results obtained, whereas criteria are basically the terms of reference used, in this case the safety of GM products for human consumption. Both of these components are integrated into risk analysis that precedes introduction of a product onto the market.

The rigorous risk analysis is comprised of toxic, allergic and composition tests. Antibiotic-resistance genes present in some of the first generation transgenic crops are being phased out. To date there are no documented cases that anyone has been harmed as a direct consequence of the use of GMO crops or products. At the same time, tens of thousands of people die yearly of food poisoning. The relative statistical danger of GM products to health is much lower than other problems facing the food industry today, a fact that is readily acknowledged by the consumer associations. A different situation is when considering a new functional food, or crops grown to be used as edible vaccines. Here, more rigorous analysis will need to be carried out to ensure that these new and different crops that substantially differ from their non-transformed parents are indeed safe to eat.

The conclusion should be to allow currently available GM products into our stores without the need of any labels (see Sect. 4.1). Indeed, on moral grounds, the money to be spent on labelling and tracing GM products could be much better spent on reducing death from food poisoning.

4.6.3.3
Scenario B: Arguments against GM Products

4.6.3.3.1
Fundamental Moral Attitudes

Genetic engineering of plants is unnatural. We should not play God; we can never do things as well as nature can or as God has provided us. Traditional plant breeding only uses what nature has given us, without resorting to technological manipulations that are man-made. Therefore we should not trust GM products. This attitude is framed mainly by religious considerations or fundamental philosophical/ideological attitudes about life.

4.6.3.3.2
Moral Attitudes that Take New Information into Consideration

No amount of testing will show that GM crops and products are 100% safe. The technology is new and the products are new. In contrast, plant breeding has been around for thousands of years and we have had time to learn and adjust to the gradual changes as new foods became available. All plants differ from each other and the comparisons between transformed and non-transformed plants are based on statistical analysis that may not reflect reality. Moreover, analysis of the food products is not rigorous enough as it may omit some important tests. The tests carried out are for the benefit of industry that may also fund some of the work done by the laboratories doing the food analysis, creating a conflict of interest. The new generation of GM crops in preparation, that have changed metabolic pathways, should be treated as medicaments, not as plants. In the end, the human population is the test ground for these new GM products.

The recommendation would be not to allow GM products in our food stores until we know much more about the safety risks. It would be unprofessional and ethically wrong to let something onto the market where we do not have all the answers.

At present, scenario A is accepted in Europe since it has been concluded that the current GM products are safe to eat. However, the answer to the question of GM food safety cannot be based only on a scientific, factual dimension of the problem. The answers given are rarely clear cut: science and knowledge is evolving. For that reason the EU has adopted a step-by-step and case-by-case approach, reviewing each new GM crop separately. It provides a safety net, just in case something should go wrong, in the form of legislations, such as the labelling and traceability regulations, that came into force in 2003. The legislation decisions taken also illustrates that EU decision makers wish to demonstrate responsibility, accountability and justification of their actions. These as already mentioned, are all components of moral and ethical dimensions.

4.6.4
Conclusion

Decisions made today are almost always based on incomplete facts. There will never be 100% safety, because that is the nature of life. We make imperfect choices in an imperfect world. That is why we need statistics to help us make decisions. Full consensus in the decision-making process is the ultimate but rarely attainable objective. The decision-making process, in addition to incorporating risk analysis, is subject to social, economic and other societal frames of reference. At the end, the decisions and actions themselves are taken within the framework of moral and ethical perspectives.

Including and deliberating public concerns as part the decision-making process will lead to co-responsibility in decision making, perhaps the best way to assure that whatever decisions we make will respect, within the rule of law, the moral and ethical standards of the society. One of the outcomes of this process should be a good practice guide: a set of bioethical guidelines for the use of genetically modified plants. Such a guide should serve as basis for creating appropriate international policies and rules that can be implemented, monitored and hopefully enforced by appropriate agencies.

4.6.5
Information Sources

Council of Europe Steering Committee on Bioethics. The committee deals with a wide range of ethical issues related to science. More information can be found at http://www.conventions.coe.int

European Commission. The European group on ethics in science and new technologies to the European Commission. http://www.europe.eu.int/comm/european_group_ethics

Food Ethics Council (1999) Novel foods: beyond Nuffield (second report of the UK Food Ethics Council). Food Ethics Council, Brighton, UK

Heaf D, Wirz J (2001) Intrinsic value and integrity of plants in the context of genetic engineering (Proceedings of an Ifgene workshop on 9–11 May 2001). Ifgene, Dornach, Switzerland. Available at: http://www.anth.org/ifgene/papersMay2001.htm

Mepham B (ed) (1996) Food ethics. Routledge, London, pp 101–119

Reiss MJ, Straughan R (1996) Improving nature? Cambridge University Press, Cambridge

5 Future Applications of GMOs

5.1
Second Generation GM Plant Products

Gert E. de Vries

Genetic traits in modified plants can be grouped into three categories:

1. Input traits, which enhance agronomic value such as herbicide tolerance, pathogen resistance or abiotic stress/yield
2. Output traits for increased quality such as modified nutrients, improved industrial use, or health-related compounds (molecular farming)
3. Traits for technological purposes such as genetic markers or gene switches

We have chosen an alternative classification following the development of GM crops in order to show the progress of this technology. A first generation of GM plants therefore received traits from the first and the third category to prove the principle of genetic modification and to acquire new agronomic traits. Second generation GM crops, to be discussed in this section, have increased agronomic and nutritional properties by inclusion of traits from both the first and second category. Third generation GM crops are modified with traits to improve their use in industrial processes, and these are discussed in the next section.

5.1.1
Introduction

The health and well-being of humans are entirely dependent on plant foods either directly, or indirectly when used as feed for animals. Plant foods provide almost all essential vitamins and minerals as well as a number of other health-promoting phytochemicals. Plants also provide the primary source of energy in the form of carbohydrates or lipids and the building blocks for proteins.

The technology of genetic modification has been used initially to produce a variety of crop plants with distinct traits to enhance their agronomic properties, such as the resistance to insect pests or viruses and the tolerance to specific herbicides. While this first generation of GM plants has benefits for farmers, it is more difficult for the consumers to see any benefit other than, in limited cases, a possible decreased price owing to reduced cost of production. This may be one of the most important reasons why today's transgenic plants are under permanent fire of criticism. This situation may change in the near future when transgenic crops of the so-called second generation enter the market. A key to the success of such GM plants is metabolic engineering: the in vivo modification of cellular processes in specific plant cells resulting in the production of modified and/or increased amounts of non-protein products. The creation of these novel crops is a more ambitious and technically challenging task since it involves the modification of plant physiology and biochemistry rather than the overproduction of a single protein.

5.1.1.1
What are Second Generation GM Crops?

Genomics-based strategies for gene discovery, coupled with high-throughput transformation methods and miniaturised, automated analytical and functionality assays, have accelerated the identification of valuable genes. The discovered genes may change the compositions of carbohydrates or the functionality of proteins and radically modify the properties of crop plants. Therefore, the real potential of GM technology to help address some of the most serious concerns of world agriculture has only recently begun to be explored. Modified crops resulting from plant biotechnology have the potential of providing major health benefits to people throughout the world. Examples include enhancing the vitamin and mineral content of staple foods, eliminating common food allergens, developing higher protein quality and quantity in widely consumed crops. The trend for this second generation of GM plants therefore is the development of GM foods with enriched nutrients, with improved functionality, and with health-promoting activities, which will all be of direct benefit to the consumer.

The third and so far the most recent generation in plant biotechnology is characterised by the use of GM plants as biofactories for the production of specialty (such as vaccines or other pharmaceuticals) or bulk products that cannot be produced in conventional crops (such as industrial oils) These developments are discussed in Sect. 5.2.

5.1.2
Examples

Detailed examples are given from an area that may appeal to consumers most – health. The development of golden rice has met a lot of debate and is therefore discussed in greater detail. The wide range of traits which are employed or considered in GM plants preclude extensive discussion here, but a useful overview is given in the first reference of under "Essential Sources".

5.1.2.1
Processing: the FlavrSavr Tomato

The first GM whole product, marketed under the brand McGregor, was introduced to the US market by Calgene in May 1994. It is interesting to note that it was in fact a second generation GM plant, producing tomatoes with delayed ripening properties. A tomato gene had been silenced that normally causes the breakdown of pectin in cell walls, resulting in a softening and eventual rotting of the fruit. This was thought to be a useful trait for two reasons. Tomatoes could be picked ripe, transported and stored for up to 10 days, providing plenty of time for shipping and sale. Secondly it was expected and advertised that the FlavrSavr tomatoes, due to the arrested loss of structure, would develop a superior taste. This was not perceived so by consumers and the first GM tomato did not meet its expected commercial success. Nevertheless, research efforts on improving the taste of tomatoes is continuing, although the majority of GM tomatoes are currently processed to ketchup and tomato paste.

5.1.2.2
Oils

Fats and oils, in particular vegetable oil sources, have been a topic of keen research interest over the past 20 years. The role of dietary fats in human nutrition has created widespread interest among consumers, clinicians and food producers. Concern with the type of fat, as an important dietary risk factor in coronary heart disease, has been a major impetus for the development of specialty modified fats and oils from plants. Consumers want oils that are low in saturated fats and low in trans-fatty acids. The food industry also requires that the oils have a high stability to oxidative changes as these can result in off-odours. The natural health products market for the omega-3 fatty acid and α-linolenic acid, found to be effective in lowering blood cholesterol levels and reducing the clotting of blood platelets and lowering blood pressure, is well established. Opportunities also exist to produce elevated levels of nutraceutical products such as vitamin E (the antioxidant α-tocopherol) in oils from plant sources. New findings in diet formulations and health-promoting substances will certainly drive further research and breeding of specialty crops that will be suited to fill such needs. A detailed knowledge of the chemistry of lipids in plant cells is an essential starting point for these studies and the identification of candidate genes with novel functions is a next step.

Oil crops are one of the most valuable traded agricultural commodities and are probably worth over 100 billion €/year. Despite the large volume of globally traded vegetable oil, only four major crops (soybean, oil palm, rapeseed and sunflower) contribute about 75% of this production. The vast majority of vegetable oils are currently used for edible commodities, such as margarines, cooking oils, and processed foods. Only about 15% of production goes towards the manufacture of oleochemicals, i.e. industrial commodities derived from oil crops (a third generation GM product, see Sect. 5.2).

In addition to the technical problems of producing useful GM oil-producing crops for all these different markets, there are considerable challenges involved in the management of such crops. Since all resulting cultivars will appear identical and the only differences are in their seed oil compositions, segregation and identity preservation of such incompatible commodity streams will turn out to be essential.

5.1.2.3
Starch

Starch is a polymer, or chain, of glucose molecules containing both amylose and amylopectin. Amylose is the straight-chain form of this polymer, while amylopectin is the branched form. Starch is the main carbohydrate food reserve in plant seeds and tubers and while it forms an important part of human nutrition, it also provides a useful raw material for industry. Amylopectin has unique physicochemical properties that makes it attractive for a vast range of non-food purposes. Most industrial uses normally involve modification by physical, chemical or enzymatic methods to alter its properties for specific purposes. There is an advantage, however, in producing a wider range of natural starches or derivatives to circumvent the need for processing steps since this is expensive and some constitute an environmental load. Transgenic potatoes and maize producing amylose-free starch have indeed been developed and are ready to enter the market. Even the creation of transgenic plants with more complex genetic changes, which would result in the production of bio-degradable plastics (a third generation GM plant product) instead of starch, are well underway. Transgenic potatoes have also been modified to prevent starch degradation resulting in sugar formation during cold storage to prevent browning (due to the presence of reducing sugars) of french fries and chips. Other potatoes with a modified high-density starch absorb less oil when deep-fried, resulting in chips containing less fat. These and other lines of research have the potential of producing healthier foods or increasing the role of agriculture in replacing limited natural resources. Again, the design of such GM plants requires a detailed understanding of the chemistry of starch synthesis in plant cells and an extensive search for suitable traits that will not disturb the physiology of the host plant.

5.1.2.4
Yield

Rice, a staple food crop for nearly half of the world's six billion population, is a key target for molecular research aiming to develop improved varieties to feed the world's expanding population. The rapid population growth, especially in developing countries, has caught up with the achieved advances in cereal yields in the past decades. To meet expected demands, according to the International Rice Research Institute experts, a 40% increase in rice yield is needed by 2020. Researchers have therefore turned their attention back to basis of crop yields–carbon fixation.

Photosynthesis is the process in green plants that uses the energy of sunlight to convert carbon dioxide and water into glucose, while releasing oxygen to the atmosphere. Glucose and derived sugars are essential for plant growth and the build up of storage compounds such as starch and oils. All animals, including humans, depend either directly or indirectly on photosynthesis, the most important chemical process on Earth. Increasing the efficiency of photosynthesis and carbon metabolism in crop plants would increase yield in agriculture and is therefore an important way to help feed the world's ever-growing population.

The process of photosynthesis is not similar in all organisms, and rice happens to employ a less efficient system of carbon dioxide fixation. Indeed, by introducing two specific maize genes that are involved in photosynthesis into rice, it was demonstrated that the new rice strains could boost photosynthesis and grain yield by up to 35% under laboratory conditions. Trials are underway for rice plants with a further enhancement of the photosynthetic capacity, requiring simultaneous expression of three key enzymes in proper cellular compartments. It is expected that the technology can also be applied to other crops with the less efficient type of photosynthesis, such as wheat.

Even plants with the efficient type of photosynthesis may be target for improvement. GM plants are being developed that need less light by increasing the amount of chlorophyll, the protein that grabs the energy quanta from light. Other opportunities to increase agricultural yield, not necessarily requiring genetic modification, can be found in improvements that influence the growing season or harvesting methods. One particular observation is worth noticing: in the Philippines a rice variety has been developed that does not die at the end of one season. Moreover, it can be grown in virtual hedgerows across mountain slopes. This perennial rice plant can be harvested again and again, while providing living barriers to soil movement on sloping land. This would help reduce loss of precious topsoil in upland regions, the silting of rivers and irrigation systems downstream, and the intensive inputs and hard labour of annual rice crops. While some of the plants have also survived experimental drought stress and yield more grain than expected, further breeding and multiplication may take five years or more before these plants are available to farmers.

5.1.2.5
Health

It is common knowledge that plants form an immensely rich source of health-promoting substances, but it is often difficult to pinpoint the true effectors, especially if combinations of substances are doing the job. Differences in requirements may exist among individuals, at what level active substances may be found in different crop varieties, how storage affects their functioning, etc. Vitamins, while essential, should not be overdosed and the recommended daily allowance (RDA) gives an indication of an average safe level. Also, there are many plants that harbour less healthy or even toxic compounds. While we have great working experience with the plants we use for food, the attention that is being given to less desirable compounds such as allergens has never been so great since the advent of genetic modification and its application to the production of food. Other health-related compounds include the production of antibodies, pharmaceutical proteins and/or (edible) vaccines. This agronomic practice, not expected for another 10 years, is often referred to as molecular farming.

5.1.2.5.1
Allergy

Our knowledge of food allergies is far from complete. It is still unclear, for example, why only certain individuals are affected and why, even among them, the problem is often restricted to childhood. It is also not clear why the allergies caused by various nuts and aquatic animals tend to persist and be lifelong. Milk, egg, soy, and wheat are the major causes of food allergies in children, whereas peanut, tree nuts, shellfish, and fish are the most prevalent causes for allergies in adults.

Virtually all allergens are proteins and food allergens typically represent 1% or more of the total protein. The question of whether a novel (transgene) protein would render a food product more allergenic than its conventional counterpart can be addressed by:

1. Comparing the predicted amino acid sequence of the novel protein with that of known food allergens
2. Examining the protein for characteristics often associated with known food allergens, such as sugar-like side-chains and heat stability
3. Monitoring the digestibility of the novel protein

Since any outcome from these comparisons would not be conclusive, a (ideally, animal) model test system would be needed to show potential evidence of allergenicity. Such a system could also be used to answer questions such as whether it is possible that the process of genetic modification renders existing proteins allergenic, or even gives rise to unintended protein products with toxic properties.

In quite a reverse line of thought the question can be asked whether genetic modification might be able to remove allergens from food, yielding for example non-allergenic cereals, dairy products and seafood. Indeed researchers in Japan are developing a variety of rice that is allergen-free and US colleagues are working on peanuts to get rid of the two main allegenic proteins responsible for a potentially life-threatening swelling of the lips and airways in sensitive individuals (estimated to affect approximately 1% of the Western population). It is not unreasonable to expect that basic research and a good understanding of the chemical processes in plant cells will deliver maize, soybean and rapeseed with a healthier fat profile and cereal grains with better nutritional value but low in measurable toxins or allergens.

5.1.2.5.2
Plant Nutraceuticals

World consumption of natural health products, nutraceuticals, and functional foods is estimated to be between $140 and $250 billion and growing at an annual rate of about 10%. Nutraceuticals, also referred to as functional foods, are ordinary foods that have components or ingredients incorporated into them to give them a specific medical or physiological benefit, other than a purely nutritional effect. A good example comes from the so-called French paradox: Epidemiological data prove that France has a low morbidity through cardiovascular coronaries (infarctus) despite the fact the diet is rich in lipids and risk factors in arteriosclerosis (think of the French cheeses and cigarettes). Polyphenols in red wine are now thought to play a role in the prevention of heart disease. Polyphenols inhibit the production of a peptide by blood vessel cells, a small protein called endothelin-1.

Endothelin causes blood vessels to constrict and overproduction of this protein compound is thought to be a key factor in why arteries clog with fatty deposits. Genetic modification may thus find ways to overproduce polyphenols, and possibly other compounds. Crops with nutraceuticals may thus help in reducing chances of developing specific diseases.

5.1.2.5.3
Vitamins: Golden Rice

A vast range of scientific studies are concerned with the role of enhanced levels of vitamins and trace elements. Vitamin E, zinc and selenium are suggest to lower the risk of heart disease and possibly improve immunity, as well as fight against Alzheimer's disease. A consensus still seems far away and a healthy, diverse diet would probably be the best choice. However, it is quite possible that certain individuals, or population groups, would benefit from foods with health-promoting elements or defined formulations. As golden rice has been such a debated subject in this respect, it useful to expand on the history of its creation and future.

Millions of people in the world suffer from vitamin A deficiency (VAD), which leads to vision impairment and increased susceptibility to diarrhoea, respiratory diseases, and measles. In Southeast Asia it is estimated that five million children develop xerophthalmia (alteration in the structure of the conjunctiva and cornea found predominantly in children) every year. This problem may be equally severe in certain areas of Africa, Latin America and the Caribbean. Overall, around 500,000 children annually become irreversibly blind as a result of VAD.

Because rice accounts for the majority of the world's diet, it was only natural that it became the focus of an intense research effort that began in 1982. Plants are usually rich in the provitamin A precursor (beta-carotene or other carotenoids) that the body converts into two vitamin A molecules (retinol), an organic compound that is soluble in fats. In the rice crop it resides mainly in the greener parts of the plant, which are discarded, and not in the endosperm, i.e. the part of the rice grain that remains after it has been polished. Traditional breeding methods have been unsuccessful in producing crops containing a high vitamin A precursor concentration and most national authorities therefore rely on complicated supplementation programs to address the problem.

By inserting three foreign genes – phytoene synthase (from daffodil), phytoene desaturase (from the bacterium *Erwinia uredovora*), lycopene beta-cyclase (from daffodil) – Dr. Ingo Potrykus of the Swiss Federal Institute of Technology and Dr. Peter Beyer of the University of Freiburg in Germany managed to engineer a full biosynthetic pathway for beta-carotene into Taipei 309, a japonica rice variety. In August 1999, they unveiled the fruit of their research and named it "golden rice." because of its colour. Golden rice varieties can easily be integrated into the farming systems of the world's poorer regions, therefore the technology promises to help solving Asia's vitamin A deficiency problem in an effective, inexpensive, and sustainable way. But how effective is golden rice really? How much carotene is needed and how much can be supplied through rice?

The FAO/WHO recommended daily amount (RDA) for vitamin A in children aged 1–6 is 400 µg. To be active, beta-carotene -(a pro-vitamin) must be split by an enzyme in the intestinal mucosa or liver into two molecules of vitamin A. Like vitamin A, the pro-vitamin is fat-soluble and requires dietary fat for absorption. Thus, digestion, absorption, and transport of beta-carotene require a functional digestive tract adequate energy, protein and fat in the diet. Therefore the RDA for carotene in healthy persons has been set to 2.4 mg. The current varieties of golden rice contain approximately 1.6 μg of beta-carotene/g dry weight. Using the standard beta-carotene-to-vitamin A conversion, 30 g of current golden rice would contain 8 μg of vitamin A activity, less than 1% of the RDA. Many children exhibiting symptoms of vitamin A deficiency suffer from generalised protein-energy malnutrition and intestinal infections that interfere with the absorption of beta-carotene or its conversion to vitamin A. Therefore, at first glance, golden rice does not seem to contribute a great deal. On the other hand, RDA values are to some extent luxurious recommendations, representing a "nice to have"-supply, which also considers the multiple effects of vitamin A and especially of provitamin A, beta-carotene. The latter, besides being a provitamin, has one additional effect, which is to act as a free radical scavenger, thus preventing typical diseases in developed countries (diseases of the cardio-vascular system and some sorts of cancer). There is consent that the amounts required in the prevention of those severe symptoms of vitamin A deficiency in developing countries are significantly lower than given by RDA values. Furthermore, it is expected that future varieties of golden rice will contain increased levels of carotenes – carrots accumulate approximately 100 μg or more of beta-carotene/g fresh weight! We will therefore know a correct answer only after having data from the varieties bred by the plant breeders, from bioavailability studies and from nutritional studies on vitamin A-deficient people.

What about patent rights and the costs of using the GM rice varieties? Will golden rice be available to farmers and thus the population of Third World countries?

The inventors, Potrykus and Beyer, were determined to make the technology freely available and since only public funding was involved they initially considered this should not be too difficult. However, at least fifteen technology property (TP) components went into the three different genetic constructs that were used to generate golden rice. Many of these TPs were acquired by ETH-Zürich under Material Transfer Agreements or by the use of licenses. Transfer and use of golden rice therefore, depending on the country in which it is to be deployed, would, at a minimum, require agreements from a dozen or so entities (public and private, institutes or companies) for the TP transfer and use. In addition, again depending on the country of use, between zero and 40 licenses for IP rights would be required, from a dozen or so entities. However, thanks to the public pressure it turned out that there was a lot of goodwill in the leading companies to come to an agreement on the use of IPR/TPR for humanitarian use that does not interfere with commercial interests of the companies. Therefore, the delivery of golden rice from the inventors' laboratories in Europe was possible as a result of the donation of intellectual property licences from a range of legal entities, including companies such as Syngenta, Bayer and Monsanto. Each company has licensed free-of-charge technology used in the research that led to the invention. Subject to further research, initially in the developing countries of Asia, as well as local regulatory clearances, golden rice can thus be made available free-of-charge for humanitarian uses in any developing nation.

To date, golden rice is a popular case supported by the scientific community and official developmental aid institutions, but equally strongly opposed by the opponents of GMOs. The first groups think it is an excellent example of how genetic engineering of plants can be of direct benefit to the consumer, especially the poor and the disadvantaged in developing countries, where GMOs offer many more opportunities for the improvement of livelihood than for those living in well-fed developed nations. The opposition, however, is concerned that golden rice will be a kind of Trojan horse, opening the door to other GM applications and leading to improving acceptance of GM food.

It is argued that vitamin A deficiency is accompanied by deficiencies in iron, iodine and a host of micronutrients, all of which come from the substitution of a traditionally varied diet with one based on the monoculture crops of the Green Revolution. Poor people do not eat plain rice out of choice, they just do not get enough to eat and are undernourished as well as malnourished. In numerous countries where vitamin A deficiency is endemic, food sources of beta-carotene are plentiful but are believed inappropriate for young children, are not cooked sufficiently to be digestible, or are not accompanied by enough dietary fat to permit absorption. In addition to doubts about cost and acceptability, biological, cultural, and dietary factors act as barriers to the use of beta-carotene, which explains why injections or supplements of pre-formed vitamin A are preferred as interventions. Opponents therefore reject the notion that golden rice could play a significant role to alleviate the VAD problems and dismiss the enthusiasm by others as a publicity campaign for the application of genetic modification.

5.1.3
Information Sources

Agricultural molecular biology laboratory (AgMoBiol) of Peking University has set up a database of food allergens. http://ambl.lsc.pku.edu.cn

Global status of approved genetically modified plants. Agriculture & Biotechnology Strategies (Canada) http://64.26.159.139/dbase.php

Information systems for biotechnology lists databases of international field tests of GMOs and commercialised GMOs http://www.nbiap.vt.edu

ISAAA brief on GM rice: Will this lead the way for global acceptance of GM crop technology? http://www.isaaa.org/Publications/Downloads/Briefs%2028.pdf

James C (2002) Global status of commercialised transgenic crops: ISAAA http://www.isaaa.org/Press_release/GMUpdate2002.htm

Lheureux et al. Review of GMOs under research and development and in the pipeline in Europe. IPTS ftp://ftp.jrc.es/pub/EURdoc/eur20394en.pdf

Prototype database for products derived using modern biotechnology. OECD http://web-domino1.oecd.org/ehs/bioprod.nsf

5.2
Third Generation of GM Plants: Biofactories

Anne-Katrin Bock, Janusz Zimny

Plants are known as sources for a variety of different pharmaceutically or industrially used substances. Modern biotechnology makes it possible through the targeted modification of plants to enlarge the spectrum of products and to generate new biofactories. It has been found that GM plants can produce vaccines and antibodies protecting against human and animal diseases such as cholera, cancer, diarrhoea or dental caries. The application also includes industrial raw materials that in future might replace petroleum-based substances. The potentials of future utilisation of products obtained from GM plants are discussed in this section.

5.2.1
Introduction

The application of modern biotechnological techniques in plant breeding promises new advances in the production of food as well as non-food products. Continuous development in these areas of research has so far resulted in three "generations" of GM plants. The first GM plants contained genes responsible for agronomically important traits, such as herbicide tolerance or insect and disease resistance, which provide advantages for farmers by reducing the necessary input of herbicides, pesticides or manpower. The second generation of GM plants are those with improved end-use quality traits (see previous section). These plants are supposed to provide improved (food) products for consumers. The introduced genes change the composition of carbohydrates, fats or the functionality of proteins. Improved amino acid composition or vitamin content (for example golden rice, a GM rice with enhanced levels of pro-vitamin A) could enhance food quality (nutraceuticals). An increase in shelf life could lead to fresher products in supermarkets and to less spoilage during transport. Also, the taste of fruits and vegetables and the decrease of allergen content in food can be influenced by genetic modification.

The third, and so far the most recent, generation of GM plants are those to be used as biofactories producing non-food fine chemicals or raw materials. Researchers may choose genetic modification of plants, but micro-organisms or animals can be attractive hosts as well. Substances to be used for diagnosing diseases or for therapy have already entered the market place. Researchers are also looking for new ways of immunising people and animals against viral and bacterial diseases. Apart from medical applications, GM plants with altered patterns of fatty acid contents find uses in the production of technical oils. GM plants may also be used to produce biopolymers replacing petroleum-based compounds in the plastics industry.

Well over a 100 compounds directly derived from plants are currently used by industry to produce pharmaceuticals. Many investigations have been carried out in recent years to improve plants as a source for such organic substances. The production of specific secondary metabolites from plants or tissue or cell culture has become routine, and genetic transformation methods have been used to increase the production of such substances.

5.2.2
Plants as Biofactories

Novel biocompounds produced by or extracted from GM plants can be used as pharmaceuticals, e.g. edible vaccines, specific antibodies or human proteins. While monoclonal antibodies, blood proteins or hormones are currently produced using biotechnological methods using animal or human tissue cultures, plants have a distinct advantage. The use of substances from human or animal sources carries a risk of contamination with pathogens while the use of GM plants would practically eliminate such risks. Furthermore, large-scale production of biocompounds from plants may be easier and less expensive.

5.2.2.1
Plant-Derived Injectable and Edible Vaccines

The immune system of humans and animals reacts against viral and bacterial pathogens by producing antibodies that are specific for certain proteins of these pathogens (immunogenic proteins). Antibodies are able to recognise these proteins, attach to them and thus initiate the destruction of the pathogenic organism. To avoid an outbreak of a known disease among human or animal populations, vaccination can be used as a precaution. Conventional vaccines usually consist of a killed or weakened form of the targeted pathogen and are usually administered by injection. The subsequent immune response will secure immunity against the disease for a defined period of time. Vaccines are currently produced by infecting human or animal cells in appropriate culture media.

Modern biotechnological methods, especially genetic modification, provide powerful possibilities for the development of new strategies for vaccine production. GM plants may serve as biofactories producing vaccines free from any potential human or animal proteins or diseases. Transgenic plants, when supplied with specifically selected genes, are capable of producing immunogenic proteins. Such plants can subsequently be grown on a large scale and can be used for isolation of pure vaccines in large quantities.

Three GM vaccines have currently been approved for commercialisation in the EU. The genetic method that was chosen relied on the production of a weakened versions of a virus after deletion of specific genes (such as the animal vaccine against Aujezky's disease) or rely on the production of a harmless host virus containing a few specific genes from a pathogenic virus (such as vaccine for foxes against rabies). While these particular vaccines are not produced in GM plants as yet, a range of vaccines from GM plant sources has been developed and produced in research laboratories.

An important factor for an effective vaccination is that the vaccine contains a protein that is specific for a certain pathogen and that the immune system will subsequently recognise the live pathogen. It was found that particular proteins provoke strong responses from the immune system when taken orally. Therefore, when searching for new forms of immunisation of humans and animals to viral and bacterial diseases, researchers have identified pathogen proteins with such properties for developing edible vaccines from plants. In order to achieve this, genes coding for such immunogenic viral or bacterial proteins have been introduced and expressed in plants. Oral delivery of vaccines is a very attractive alternative to injection. Production in GM plants is comparatively inexpensive, it can be done in any country, including developing countries, purification of the vaccine is not necessary and administration is easy.

So far research has predominantly been carried out with plants that are easily modified, such as tobacco, potato and tomato. After testing the reliability of the system, other plants will be used that can be eaten raw or that are suitable for storage as such as banana, lettuce, wheat, rice and maize. Therefore, oral vaccinations against common diseases such as flu, hepatitis B, tuberculosis, malaria and cholera might indeed become a reality in the future.

5.2.2.1.1
Examples of Vaccines Produced in Plants

Several strategies to produce effective vaccines in plants have been developed. One of them makes use of a plant virus, harbouring the gene for a protein of a human virus. After infection of a plant with this GM plant virus, the human virus protein is produced and can be isolated, purified and used as a vaccine. Indeed, animals, when vaccinated, developed antibodies that would react with the human virus. Vaccines for the human immunodeficiency virus (HIV) and the human rhino virus (HRV) have been developed following this method.

In potatoes, an orally active vaccine against cholera has been developed. Cholera-causing bacteria produce two types of toxin proteins but only one of them is toxic for humans. The harmless protein can therefore be used to induce immunity. A part of this protein was produced in potatoes after introducing a gene fragment coding for the toxin. Mice that were fed with the GM potatoes, were found to produce antibodies that would neutralise the activity of both cholera toxins.

Another important target for vaccine production is hepatitis B, a virus that is responsible for the majority of chronic liver diseases. Vaccines that are effective against hepatitis B infection have been produced in yeast, but these are expensive to produce and need stringent storage conditions. Therefore novel vaccines have been developed using potato and lettuce plants. Mice fed with GM potatoes producing one of the hepatitis B virus proteins showed an immune response. Future efforts will include banana as a vaccine source since such a pharmaceutical crop can easily be grown and administered in developing countries.

The Norwalk virus is responsible for most non-bacterial gastroenteritis. A protein that forms the coat of the virus, the capsid, has been expressed in transgenic tobacco and potatoes. After injection of these purified antigenic proteins into mice, protein-specific antibodies were produced.

Not only fruit or leaf tissue but also plant seeds offer potential advantages for the production of vaccines that can be administered orally. For example, a glyco-

Table 5.1. Examples for vaccine production

Expressed protein	GM plant host
Heat-labile toxin enterotoxic *E. coli* (humans)	Tobacco, potato, maize
Cholera toxin of Vibrio cholerae (humans)	Potato
Envelope surface protein hepatitis B virus (humans)	Tobacco, potato, lettuce
Capsid protein of Norwalk virus (humans)	Tobacco, potato
VP1 protein of foot-and-mouth disease virus (agricultural domestic animals	Arabidopsis, alfalfa
Glycoprotein S from transmissible gastroenteritis coronavirus (pigs)	Arabidopsis, tobacco, maize
Epitope gp120 and gp41 human immune deficiency virus (humans)	Tobacco, black-eyed bean, cow-pea
Epitope protein of human rhinovirus	Black-eyed bean
Glycoprotein B of human cytomegalovirus	Tobacco

protein of the human cytomegalovirus (a widespread virus, that causes pneumonia in immune-suppressed patients) that was expressed in GM tobacco seeds proved to be immunologically reactive.

Immunisation by edible vaccines of animals also has an economically attractive potential. For instance, vaccines active against a transmissible gastroenteritis virus of pigs, when produced in maize, were successful in protection against the disease (Table 5.1).

5.2.2.2
Therapeutic and Diagnostic Antibodies

An alternative to immunisation, and thus to relying on antibody production in the patients, is the direct administration of appropriate antibodies to treat a disease (passive immunisation). Such antibodies can also be used for diagnostic testing. The ability of plants to produce functional monoclonal antibodies, so-called plantibodies opens the way for the establishment of an efficient and inexpensive method of immune protection against a number of diseases.

Only a limited number of antibodies that could be used in human medical applications have so far been produced in plants. Antibodies against the bacterium *Streptococcus sanguis* have been obtained from GM tobacco. Oral application of these plantibodies in clinical trials showed prevention of the formation of tartar on the teeth and, consequently, of dental caries. Plantibodies that inactivate *Streptococcus mutans*, another caries-causing bacterium, are at the stage of clinical testing. The expression of monoclonal antibodies against the herpes virus, causing fever blisters, has been reported for GM soybean. Rice and wheat plantibodies against a cell surface protein typical of tumours (carcinoembryonic antigen) have been produced and can be used for diagnostic purposes as well as in tumour therapy.

Table 5.2. Examples of plantibody production

Application for the plantibody	GM plant host
Dental caries (against streptococci)	Tobacco
Cancer treatment/diagnosis (against carcinoembryonic antigen)	Wheat, rice, tobacco
Herpes treatment (against herpes virus)	soybean

Because of the production in plants, plantibodies show slight modifications in molecular composition as compared to human antibodies. Although these differences do not influence their effectiveness, there is the potential of an allergic or even immunogenic reaction. Novel human therapeutics must be carefully checked and for this reason plantibodies are not yet produced commercially. Although it is expected that production costs from plants will be much lower, the cost of purification will still be comparable to current practices. But, purification might be avoided if antibodies were produced in seeds and oral application proved effective (Table 5.2).

5.2.2.3
Other Biopharmaceuticals

Apart from vaccines and human antibodies, other human proteins or pharmaceuticals may also be produced using GM plants. Several GM plants have already been developed for the production of, e.g. interferon, human serum albumin, haemoglobin, blood coagulation factors etc. (see Table 5.3).

Human haemoglobin as well as blood coagulation factors have been synthesised in GM tobacco plants. Glucocerebrosidase, an enzyme that is deficient in patients with Gaucher's disease, is an expensive drug since 10–12 tons of human placentas are used to isolate and purify the enzyme each year for a single patient. The enzyme has now also been obtained from GM tobacco, where it is expressed after harvesting the plants in order to limit accidental environmental exposure to pharmaceuticals.

Hirudin is a protein that is obtained from leeches and is used for its anticoagulant activity in the treatment of thrombosis. GM oilseed rape that produces

Table 5.3. Examples for protein/biopharmaceutical production

Application	GM plant host
Anticoagulant (hirudin)	Oilseed rape
Hepatitis B and C treatment (interferon)	Rice, turnip, tobacco
Blood substitute (human haemoglobin)	Tobacco
Gaucher's disease (glucocerebrosidase)	Tobacco
Wound repair, control of cell proliferation (human epidermal growth factor)	Tobacco, kiwi, potato, orange
Liver cirrhosis, burns, surgery (human serum albumin)	Tobacco

hirudin is already cultivated and commercialised in Canada. The plant produces a hirudin-oleosin fusion protein, a trick which results in storage of the hirudin protein in seed oil-bodies, thereby facilitating purification. Hirudin is recovered after enzymatic cleavage of the fusion protein. Apart from an improved purification procedure, the chosen approach also leads to hirudin only being activated after harvest and purification, thus limiting potential environmental impacts.

The gene of a mussel glue protein has also been introduced into plants. The water-resistant adhesive, which has a breaking load twice that of most epoxy resins, does not attack human cells or provoke an immune response, therefore it could be ideal for repairing soft tissues and bones inside the human body. Finally, expression of the human epidermal growth factor was reported in GM kiwifruit, potato and trifoliate orange. The application of these and other human growth factors may be useful to support the repair of tissues such as skin, bones or nerves (Table 5.3).

5.2.2.4
Biopolymers from Plants

The manufacture of plastics and polymers from petroleum-derived chemicals has a negative environmental impact, apart from the fact that this practice relies on the use of non-renewable resources. Plants would be a much more acceptable source of biodegradable polymers for the plastics industry. Genes responsible for the production of a certain type of polymer were identified in bacteria, isolated and introduced into mustard thale cress (*Arabidopsis thaliana*) and maize plants. Maize was modified in such a way that the required polymer was produced in leaves and stems but not in grains. Thus, the grains could potentially continued to be used as food, and the rest of the plant as a resource for the chemical industry. However, thus far the production of plastics from GM maize is still too costly and therefore the process has not been commercialised. Nevertheless, field trials are underway with oilseed rape and soybean as well.

Spider silk is known to be a very strong and flexible fiber. Increasing knowledge about its structure and the identification of the genes that code for the proteins involved facilitated the production of spider silk in GM plants ("biosteel"). Applications include fishing lines, ultra-light protective vests or stitching materials in the medical field. While tobacco and potato plants were modified to produce the spider silk proteins, success was also reported when GM mammary cells in cows and goats produced the proteins in milk and the proteins could actually be spun into fibres.

5.2.2.5
Oils for Food Production and Technical Use

The possibility of the genetic modification of the oil content and fatty acid composition in plants and seeds can play an important role in the production of healthier foods as well as in finding new or improved materials for industrial use. A large proportion of this type of research focuses on oilseed rape, the third most important oil-producing crop in the world. GM varieties with high levels of stearic acid or high myristic-palmitic acid as well as high erucic acid content have been developed. A variety with high levels of laureate oil is available commercially under the brand name Laurical. Its oil products are used in coffee creamers, in whipped toppings and in the confection industry. GM soybean and sunflower varieties gain interest because of the high quantities of oleic acid in these plants, which are important in the food industry.

There are also expectations for vast new markets for non-food bio-oils. World crude oil production from petroleum reserves will probably peak at some time between the years 2000 and 2020. Crops like sorghum, soybean and rapeseed are being developed and used for the production of fuels like ethanol, diesel petroleum or for fuel additives and lubricants.

New vegetable-based oils can also serve as a source for a wide range of biodegradable petrochemicals, which are the raw materials for products such as plastics, textiles, lubricants, paints and varnishes. Once non-renewable hydrocarbon resources such as petroleum and coal are exhausted, there will be no other source of such products. An example is oleic acid that can be converted into estolides to be used in production of hydraulic fluids. Researchers have also developed transgenic plants expressing enzymes that convert oleic acid into vernolic or crepenynic acids for the production of environmentally friendly paints.

5.2.2.6
Production Methods

5.2.2.6.1
Field-Grown Plants

GM crops can easily be grown in the field to produce specialty or bulk substances on a large scale. It must be kept in mind, however, that non-food producing crops may need to be segregated from other crops. Pharmaceutical substances may be quite toxic in large doses and therefore GM crops for these purposes should be stringently controlled. GM crops for the production of industrial bulk compounds should not be allowed to mix with food or feed crops. Indeed such segregation is already standard practice for (non-GM) rapeseed crops with enhanced levels of erucic acid, which has anti-nutritional characteristics and is used as an industrial raw material for, e.g. lubricant production.

The possible persistence of novel plant products in the environment and their possible effects on non-target organisms must also be examined carefully. Out-crossing to other crops or related weeds may be undesirable and must therefore be prevented. Therefore, depending on the nature of the product and the characteristics of the crop, the degree of containment needs to be decided upon. It may be necessary to employ genetically sterile plants, restrict the activation of the desired product to the laboratory setting or to physically confine growth and harvest in specially equipped greenhouses.

5.2.2.6.2
Plant Cell Culture

An alternative to crops that are grown and harvested from an open field or in greenhouses might be the use of suspension plant cell cultures. Plant cells can be multiplied very efficiently in liquid media. Cell suspension cultures can be used as biofactories to produce substances of interest without any contact with the environment and without risk of mixing with feed or food. An illustration of such a set-up is the production of recombinant human interleukin using GM yeast or tobacco cells that are cultured in a suspension culture. The plant cells are able to excrete interleukins through the plasma membrane and cell wall into the medium, from which isolation and purification of the substance is relatively easy. Interleukins play a role in the regeneration of tissues, are involved in immune regulation and are used for the treatment of diseases such as asthma and cancer.

A successful method for generating large cell suspension cultures with identical genetic makeup, employs the bacterium *Agrobacterium rhizogenes*. These bacteria are able to reliably transfer a small and distinct fragment of DNA into the nucleus of a plant root cells, causing tumour-like effects and the formation of hairy roots. The plant root cells can be grown in a controlled way, using specific growth substances, and can be induced to produce specialty secondary plant metabolites such as phenolics, alkaloids and terpenoids. The amount of secondary metabolites developed in hairy root cells can be regulated by certain supplements added to an in-vitro culture medium (e.g. plant growth hormones or ammonium) or other culture conditions (e.g. light intensity, temperature).

The biosynthesis of geranin, which is used to fight diarrhea, can thus be increased in the hairy root culture of geranium (*Geranium thunbergi*). The medical plant great burnet (*Sanguisorba officinalis*) is a source of the compound sanguiin, which has hemostatic and anti-inflammable activity. Hairy roots produced twice as much sanguiin as the conventional plant. Scopolamine (used as a tranquiliser as well as for diagnostic purposes) was produced with high efficiency in the deadly nightshade (*Atropa belladonna*) root cultures after *Agrobacterium* mediated transformation.

Other pharmaceutical substances, like piperidine and lobeline, were isolated from the hairy root cultures of lobelia (*Lobelia inflata*) and henbane (*Hyoscyamus albus*). Another example is the production of glycosides, used in cardiology, in hairy root cultures of foxglove (*Digitalis purpurea*). Cancer fighting substances were isolated from the hairy root cultures of yew (*Taxus*), platycodon (*Platycodon*) and lobelia.

5.2.3
Information Sources

Arkansas Agricultural Experiment Station, University of Arkansas Division of Agriculture. Using plants as biofactories is made possible by the increasing understanding of the genetic structure of organisms. http://www.uark.edu/depts/agripub/Publications/Agnews/agnews01–47.html

Biofactories. Within the plant biotechnology sector there is great interest in expressing mammalian proteins in plants, in a way that would allow their commercial exploitation. http://www.dcwi.com/~pnpi/Biofactories.htm

iniell H et al. (2001) Medical molecular farming: production of antibodies, biopharmaceuticals and edible vaccines in plants. Trends in Plant Sci 6:219

Giddings G et al. (2000) Transgenic plants as factories for biopharmaceuticals. Nature Biotechnol 18:1151

Giddings G (2001) Transgenic plants as protein factories. Curr Opin Biotechnol 12:450

Sheller J, Gührs K-H, Grosse F, Conrad U (2001) Production of spider silk proteins in tobacco
and potato. Nature Biotechnol 19:573–577

The Canadian Food Inspection Agency organised a public forum on plant molecular farming
http://www.inspection.gc.ca/english/plaveg/bio/mf/mf_come.shtml

The Pew initiative on food and biotechnology hosted a workshop exploring the potential risks
and benefits of bioengineering plants to produce pharmaceuticals http://pewagbiotech.org/
events/0717/ConferenceReport.pdf

5.3
Biotechnology for Food and Agriculture: Global Issues

Gert E. de Vries

Current and future developments in agricultural biotechnology may contribute to prevent local food shortages and improve its quality. Technological progress will nevertheless neither be a major solution to current problems nor can it be ignored because immediate solutions are not manifest. In this section a range of issues are described that are closely connected to, or interfere with, a successful deployment of improved crops that will arise from biotechnological research and development.

5.3.1
Hunger and Poverty

The number of chronically malnourished people in the world in 2001 was estimated to be 840 million, roughly equivalent to the combined population of the United States, Canada, Russia, France, Germany, the United Kingdom and Japan. Of these people, 95% live in the developing world. There is enough food produced for the whole world community, so why is it so difficult to curb malnutrition and famine? Is food security basically a distribution problem, as often advocated? Food is handled as a mere commodity in world trade, therefore adequate distribution would require overseeing and a change in policy. As long as food is not recognised as a basic human right, improvements in agricultural technologies and the stimulation of fair trade seem to form the best options. Among reasons for the existence of hunger are:

1. Unstable governments and political unrest
2. Lack of infrastructure and distribution systems
3. The toll placed by tropical diseases and HIV on the agricultural work force
4. Increased frequency of occurrence of natural disasters like earthquakes, extensive periods of drought, hurricanes or mudslides which partially may be connected to effects of global warming

These and other effects of human intervention in the environment may become a trend and will probably intensify over the next 30 years.

The greatest problem for the inhabitants of the developing world is having no financial reserves nor technology to curb any of these mis. makes them increasingly dependent on populations in rich countries. These disparities are growing. The richest 15 persons in the world have more wealth than the combined yearly gross domestic product (GDP) of all of sub-Saharan Africa with its 550 million people. There are more numbers for reflection: the world's richest countries, with 20% of the global population, account for 86% of private consumption while the poorest 20% account for just 1.3%. A child born today in an industrialised country will add more to consumption and pollution over his or her lifetime than 30–50 children born in developing countries. Half the world, nearly three billion people, live on less than $2 a day and 30% of them are unable to read or write. To place this situation in another perspective: the costs for schools for all children equals 1% of the money spent on weapons in the world. The condition of hunger in a world of plenty is seen by some to be as equally monstrous and unconscionable as slavery has been in the past. Therefore every effort should be done to open eyes to these dramatic events which claim tens of thousands of hunger-related deaths every day.

Lack of food is an obvious condition, but deficiencies in micronutrients such as vitamin A, iodine, iron and zinc are widespread with equally serious consequences for health. The World Bank's World Development Report 1993 found that micronutrient programs were among the most cost-effective of all health interventions. This general problem of poor dietary quality has been dubbed "hidden hunger".

At present, one-third of the population of sub-Saharan Africa falls below the poverty line and, according to USAID data, an estimated 50% of the world's hungry will reside in this region in 10 years from now. Experts say that there cannot be a long-term solution to famine without significant investment in education and agriculture, which is the foundation of most economies in Africa. Farming supports more than 70% of the population and contributes about 30% of gross domestic product. However, African farmers face stiff competition: per capita world food production has grown by 25% over the past 40 years and food prices in real terms have fallen by 40%. While current practices of small scale agriculture may be more acceptable to support the local social fabric of life, it may not be an economically viable option to compete at a world scale.

5.3.2
Problems Facing World Food Production

Despite steadily falling fertility rates and family sizes, the world population is projected to grow from 5.6 billion in 1994 to up to 9 billion in 2050. The population in the 50 least-developed countries will nearly triple in size in 50 years and by then, 84% of the world population will live in what we call today developing countries.

According to the FAO more than 800 million people around the world currently do not get enough to eat. On average an adult needs 2,200 kcal/day. According to UN data, the world food supply in 1970 represented 2,360 kcal per person each day and has risen to more than 2,750 kcal today. The rate of increase in food supply is expected to exceed that of population growth until 2010, suggesting that sufficient food is available to feed the world population.

So far, predictions that the human world population would multiply beyond their capacity to feed themselves have repeatedly been proven wrong. Thomas Malthus predicted famine in 1798, just as farm yields were taking off. Similarly, Paul Ehrlich in 1969 gave up hopes of continuing to feed India and concluded that hundreds of millions of people would starve to death in the 1970s. Indeed, the world's population grew much as expected, but food output more than kept pace. The Green Revolution brought novel crop varieties that farmers rushed to adopt.

Research that was exclusively undertaken by institutions in the public sector during the years 1960–1980 introduced semi-dwarfed wheat and rice varieties to parts of Asia and Central America, together with well-functioning systems of irrigation. Millions of farmers started using higher-yielding hybrid seeds for the production of staple foods, chemical fertilisers, pesticides and weed-killers. These agronomic technologies even allowed India to export surplus grain. Chinese rice farmers also raised production by two-thirds between 1970 and 1995, but these increases largely bypassed sub-Saharan Africa. While the new technologies may still have saved a billion people from starvation in the past, the subsequent population growth, changing demographics and inadequate poverty intervention programs have now probably levelled most of its gains.

It is sometimes argued that agricultural technology merely postpones an inevitable next famine. By making more food available, the world population will grow further, leading to additional demand for agricultural lands and a further conversion of wildlife habitat. However, agricultural technology not only results in larger quantities but also enhances food quality, helping to reduce general misery, maternal and infant mortality rates, and thus directly improve human well-being. A better quality of life may very well slow down population growth, as witnessed in the developed countries.

The pledge made at the 1996 World Food Summit to reduce by half the number of hungry people in the world by 2015 is far behind schedule. Therefore, the FAO concluded that the world lacks the commitment to ensure that all of its population has enough to eat and that poor countries must do more to stimulate their own agro-economy rather than rely on foreign aid. It is indeed conflicting wisdom to note that subsidised farmers in rich countries produce enough surplus food to feed the hungry, but not at a price they can afford by regular trade. Much food is therefore donated. However, since most poor people earn their living from agriculture, quantities of donated food negatively affects their livelihoods and the build-up of market economies in the target countries.

The world's population is expected to almost double between 1990 and 2050, making food security one of the most important social issue for the next 30 years. Food production will have to be doubled or preferably tripled to meet the needs of the expected 9 billion people, 90% of whom will reside in the developing world. However, natural resources, necessary to feed an extra 250,000 world citizens/day, are declining: the availability of productive new croplands is decreasing, there is a lack of fresh water for irrigation expansion and crops show only limited response to additional fertilisers. The climatic conditions in developing countries, mainly in the tropics and subtropics, favour insect pests and disease vectors causing significant crop losses only to be worsened by poor soils, low productivity levels and post-harvest losses. Poor farmers lack economic resources to purchase high quality seeds, insecticides, and fertilisers. The only answer to world hunger, aside from necessary political and infra-structural reforms and assuming that population growth cannot be curbed, is to improve the productivity of farmlands in poor countries and support poor farmers when adopting successful technologies. In addition, there is a need for rich states to open up their markets to poor nations. Poor countries currently cannot compete fairly in agricultural markets because of a trade imbalance caused by liberal subsidies to agriculture in the countries of the Organisation for Economic Cooperation and Development, compared with marginal farm aid programs to developing nations.

Apart from improvements of political issues and resolving unstable situations, there are a number of technological options to improve food production:

1. Increase local farm productivity in the countries that most need the food
2. Work hard on new technologies to improve or modernise existing agricultural practices
3. Expand the area of farmland at the cost of forests, grasslands and risk losing areas of important biodiversity
4. Intensify production in agricultural exporting countries and distribute the products to where they are needed

There is not one single answer and any theoretically sound solution may have far reaching effects on economy, security, food quality and environment. A number of issues will be dealt with below.

5.3.2.1
Food Distribution and Aid

It is an embarrassment that in today's world, when global food production should suffice to feed everyone, so many people are malnourished or starve to death. It is therefore often argued that food security is a distribution problem rather than a production problem. Indeed if the world's food supply were equally distributed everyone could have an adequate diet.

In developed countries people prefer types of food that use lots of agricultural resources for a given level of nourishment. A serious redistribution effort would therefore require convincing about a billion people in developed countries to change their eating habits and give up certain types of food. Furthermore it can be questioned how long-term shipments of food can be financed or compensated by trade.

It is therefore not easy to imagine how such a change would take place while preserving democracy based on capitalism in food-producing countries. This culture is in many ways the most successful that has ever been deployed in terms of accommodating large numbers of individuals in comfort and luxury. However, it has not been as successful in integrating all in equal measure. While it has solved the problems of feeding large numbers of people, its failure in achieving equality on a world scale remains one of its major problems.

Solving food security by equal distribution therefore is clearly too idealistic for the current political world situation. Also, one cannot argue that there is enough food, for instance, in the continent of Africa and that it is merely a matter of distribution to prevent mass starvation. Above all, Africa has inadequate infrastructure for the transportation of food since only 20% of roads are paved, making transportation of perishable goods difficult. Africa is plagued by political unrest in different regions, which prevents food from reaching the poorest of the poor. Many of the world's poor, in fact, live in countries where governments lack either the will or the ability to raise living standards on their own. The Food and Agriculture Organisation (FAO) of the United Nations (UN) has listed wars and other forms of armed conflict as the exclusive cause of food emergencies in 10–15 developing countries during the last three years. Financial assistance to such governments, therefore, has often not helped their neediest citizens. Foreign aid has even worsened the worries of the poor by sustaining the corrupt or otherwise inefficient governments that caused their misery in the first place.

The best way, therefore, to combat this inequality is for rich countries to increase foreign aid budgets and initiate new, Marshall Plan-like, initiatives to fight poverty and improve local production systems. However, for such plans to succeed there must be an existing infrastructure where aid can flourish. It is sad to note that in the poorest regions of the world there is either little infrastructure, or existing (political) systems actually cause poverty. For instance, the future will tell how serious the consequences will be after Zimbabwe's president Mugabe ordered the confiscation of white-owned farmlands in the year 2001, thereby dismantling the existing agricultural infrastructure without providing adequate alternatives.

In the following year, US relief food packages were rejected in Zimbabwe since the unmilled maize seeds were genetically modified. It was argued that farmers would use the food as planting material and fields would be contaminated with GM maize. This would potentially render a large portion of the nation's future maize harvests to be unexportable to Europe and other nations that restrict imports of GM foods. It is likely that such problems will reoccur in other parts of the world, since GM foods continue to be produced in increasing quantities. In reaction, the USAID and the World Food Program, while not specifically referring to GM foods, claimed that they never have or will distribute foods that are not fit for human consumption or which might damage people's health in any way. However, since seeds from food aid donations are likely to be planted by poor farmers, it is not wise to introduce GM crops, against prevailing regulations, into any country by this route.

5.3.2.2
Climate, Soils, Nutrients, Water and Yields

In developing countries in the tropics and subtropics, crop losses due to pests, diseases, and poor soils are augmented by unfavourable climatic conditions such as droughts and by the lack of economic resources to purchase high quality seeds, insecticides, and fertilisers. In addition to low productivity levels there are high post-harvest losses due to the lack of appropriate storage facilities to prevent fungal and insect infestations. Indeed, pests destroy over half of all world crop production.

5.3.2.2.1
Water

While approximately 50% of the world's food production is fully dependent on timely precipitation, up to 40% of the crops are irrigated with water collected from rainfall and 10% is grown using groundwater or water from wells. There is growing competition for this water between cities and industry, with agriculture often getting the worst quality of water, if any remains. While supplies of water are generally adequate to meet demand for the foreseeable future, water is poorly distributed across countries, within countries and between seasons. Modern agriculture is by far the largest consumer of water, accounting for 70–80% of all water use. Lack of available water will seriously limit the expansion of food production when using current crop varieties. World water use has risen sixfold over the past 70 years, therefore the rapidly growing domestic and industrial demand for water will have to be met by reducing the use in the agriculture sector. A projected doubling of food production in unfavourable climates must therefore largely take place on the same land area while using dramatically less water.

Photosynthetic carbon dioxide fixation by plants is associated with a large amount of water loss through transpiration. The growth of quality crops requires considerable and varied amounts of water. For instance, the production of 1 kg of cotton by irrigated agriculture currently requires 17,000 L of water while growing 1 kg of rice requires about 5,000 L. The improvement of water delivery equipment may help to drive these figures down, since most of this water does not reach the plants but evaporates. There are also biotechnological solutions. To prevent desiccation-induced growth arrest and injury, most plants require a constant level of adequate soil moisture. However, recent advances in understanding the genetic control of drought tolerance offer new opportunities to develop crops that are less damaged by short periods of low soil moisture. This might enable the use of less water for irrigation and reduce drought-induced yield reduction caused by unfavourable climatic conditions. More effective management of water can therefore be reached through a series of institutional and managerial changes, in addition to a new generation of technical innovations that may also include the exploitation of advances in genetic modification of plants.

The costs of developing new sources of water are high and rising, and non-traditional sources such as desalination, reuse of wastewater, and water harvesting are unlikely to add much to global water availability in the near future, although they may be important in some local or regional ecosystems. A Blue Revolution, complementing the past Green Revolution, may be badly needed if we are to reform the usage of land and water while meeting required environmental and economic conditions.

5.3.2.2.2
Soil Fertility

Improved soil fertility is a critical component of increased sustainable agricultural production. Reduced use of fertilisers is warranted in some locations because of negative environmental effects. But, in countries where soil fertility is low and a large share of the population is food insecure, fertiliser use may need to be expanded. Locally available organic materials are usually insufficient by themselves and while nitrogen fixation by legumes may help, inorganic fertiliser will continue to be an important plant nutrient supplement.

Abiotic stress is a major limiting factor in agricultural crop production in many countries. The major abiotic stresses of economic importance include drought, cold, heat, salinity, soil mineral deficiency, and toxicity. Combined with chronic shortage of water, the results of poor soil fertility for African farmers is easy to predict. A study on vegetable production in South Africa showed that soil acidity and drought stress accounted for over 80% of all crops that were lost. Diseases and pests accounted for the remaining 20%.

Salinity is usually worsened by intensive irrigation. Agricultural lands that repetitively need extra water due to drought periods, or lands that experience high water losses due to evaporation run the risk of accumulating salts. Rice fields, which are mostly irrigated, are declining in productivity in many Asian countries because of increasing salinity levels. It is estimated that 25 million ha of agricultural land in the world suffers from excess salinity, and approximately 10 million ha can no longer be used for agricultural production. In 30 years, about 10% of all arable land will be contaminated with salt concentrations that will seriously affect crop yields.

Traditional plant breeding has had almost no success in creating crops with sufficient resistance to increased salt concentrations, but there is some progress in wheat breeding. Although some wheat cultivars are considered moderately tolerant to high salinity, they are still less salt-tolerant than many wild *Triticeae* species, especially those in the genus *Thinopyrum*. A long-term effort by plant breeders may lead to the transfer of salt tolerance from such species into wheat. However, genetic modification promises to be a much more powerful approach. In 2001, Chinese researchers reported to have isolated GM plants of tomato, eggplant, and hot pepper that even can be watered with seawater after introduction of genes from salt-resistant plants such as mangrove.

Soil acidity is the term used to express the quantity of hydrogen (H^+) and aluminium (Al) in soils that are highly weathered (leached) because of excessive rainfall. This process has depleted nutrient elements such as calcium (Ca^{2+}) and magnesium (Mg^{2+}) from naturally occurring minerals by chemical reactions involved in the nitrogen cycle and breakdown of organic matter. Soil acidity is a major problem for about 30% of all arable land in tropical regions. The presence of aluminium at levels exceeding 200 ppm presents a serious abiotic stress for most crops. The most important symptom of aluminium toxicity is the inhibition of root growth, caused by inhibitory effects on calcium and magnesium import at the level of the cellular membrane. Long-term exposure of plants to aluminium also inhibits shoot growth by inducing nutrient deficiencies, drought stress and phytohormone imbalances. Metal toxicity and nutrient deficiency problem acid soils are investigated by only a handful of scientists in developed countries and this topic has thus far been largely neglected by large agrochemical companies. Sub-Saharan Africa has some of the oldest and most depleted soils in the world. Thousands of years of weathering have leached the nutrients, leaving the soil highly acidic, causing aluminium and manganese to become soluble and thereby toxic to plants. Together with high levels of iron, the aluminium oxides also hinder plant growth by chemically locking up phosphates. So far, traditional plant breeding has produced few answers for this global problem.

5.3.2.3
Impact on the Environment

Human activity is altering the planet on an unprecedented scale. More people are using more resources with more intensity and leaving a bigger "footprint" on the earth than ever before. Humans have become a major force of nature, largely because of the success of science-based technologies in extracting the earth's resources without proper concern for the environmental consequences. Science, though, also has a crucial role to play in helping us avoid the impeding catastrophe that is partly of its own making. Paul Ehrlich, author of the book *The Population Bomb*, developed an equation in the 1970s to describe the impact of human population on the environment. The population-resource equation puts the challenge that faces agriculture into perspective by saying: (Natural resource use)×(technology)=(population)×(per capita consumption). It would mean that, unless plant science and agricultural technologies can bring significant progress, natural resources will be increasingly employed to feed a growing world population with rising demands.

Agriculture accounts for 38% of land use, for some 75% of water consumption and it is responsible for most of the habitat loss and fragmentation that threaten the world's forests and biodiversity. Pesticide or fertiliser runoff and soil erosion threaten aquatic and avian species from sweet and coastal waters. But, paradoxically, if agricultural technology had been frozen at 1961 levels, the level of agricultural production in 1998 would have required more than a doubling of land devoted to agriculture: an extra area approximately the size of South America. Since the best agricultural land is probably already being cultivated and new cropland is unlikely to be as productive, this may even be a conservative estimate. Thanks to improvements in productivity this could still be avoided, but, as the human population rises from 6 to 9 billion people in the next 50 years, the environment may be further stressed by necessary increases in agricultural farmlands. There are still areas, mainly in Africa, where the techniques of the Green Revolution have yet to be tried, but in most of the developing world the gains in productivity are tailing off. Perhaps again we need a new impulse from science in developing sustainable agricultural systems that will curb the trend.

5.3.3
Strategies for the Improvement of Agriculture

There have been a number of ingenious suggestions for changes in agricultural practices that would improve the environment, while at the same time increase the efficiency of fertiliser and pesticide use, thus maximising crop yield. One example is "precision agriculture" in which the state of the soil and crop is monitored, sometimes by satellite, so that sowing and treatment rates can be adjusted accordingly. Such state-of-the-art agriculture technologies require high investments and are still out of reach of most farmers today. Alternatively, easy-to-use electronic tools, such as soil nutrient detectors or leaf chlorophyll measurements, could also play an important role in the improvement of current conventional methods in agriculture. Another method is the no-till strategy, combining organic and conventional methods and dubbed "integrated agriculture", in which soil structure and biodiversity is conserved by obviating the need for ploughing and natural predators are employed to ward off plant pests.

Growing crops organically, a third strategy, means not using synthetic fertilisers and only a few selected pesticides of natural origin. Proponents argue that organic agriculture maintains a better soil quality, increases biodiversity in the field, produces equal or better food qualities and lowers the vulnerability of crops to pests. While conventional agriculture is seen to deplete resources and is susceptible to plagues or introduces unwanted chemicals in food products, organic agriculture is advocated to guarantee sustainability and to respect nature. Sustainable farm practices must nevertheless lead to adequate high-quality yields, be competitive and profitable, and in addition protect the environment, conserve resources and be socially responsible in the long term. Specific indicators used are soil quality, performance, profitability, environmental quality and energy efficiency.

Organic agriculture is not all as rosy as it seems since natural pesticides contain compounds that can be as poisonous as man-made chemicals. In addition, if farmers do not effectively treat fungal infections in their crops, mycotoxins may be produced which rank very high on the list of carcinogenic compounds. It is often questioned whether natural pesticides and organic methods will continue to sufficiently ward off plant diseases or whether a combination with biotechnological solutions should be sought.

Other problems would arise if it were decided to grow all food according to organic standards. Approximately 40% of the world population of 6 billion use crops grown using synthetic ammonia fertilisers. An extra 5 billion cattle would be required to supply the manure needed to produce the yearly 80 million tons of nitrogen nutrients used globally each year. There are no clear predictions what the consequences would be for the less precise applications of fertiliser, for hygiene, for environmental pollution due to runoff or for the increase of volatile ammonium loss in the atmosphere. Because organic crops require nitrogen obtained from supplementary farm practices rather than manufactured in a factory, organic farming is also land-hungry. If the current 6 billion people were to be fed using the technologies and crop yields of 1961, which were mainly organic, it would now require an approximate tripling of land area to be cultivated.

This assessment is, however, somewhat inadequate since organic farming has improved since then and agricultural techniques in 1961 were not fully organic.

Studies on the actual energy consumption of organic, conventional and integrated agriculture also point out that organic agriculture may use more energy and produces more carbon dioxide and ozone layer damaging nitrogen oxides. The higher energy consumption was due in part to the need for mechanical weeding and cattle-rearing activities, which easily exceeded the amount of energy needed to produce nitrate fertiliser in non-organic systems.

In developing countries organic agriculture is practiced in many places, not because of choice but because of lack of resources. Farmers in developing countries may, in comparison to large capital-intensive farms in developed countries, use inefficient methods but these are directly responsible for preserving vast amounts of plant biological diversity. Developing countries possess the largest plant biological diversity on earth, as well as the largest problems of soil depletion and environmental degradation. It is likely that agricultural production will increasingly specialise, with exports from countries that focus on particular products and types of agriculture. Many developing countries may well hold a comparative advantage in producing high-value, labour-intensive specialty crops and horticulture, while land-abundant countries may be better at producing bulk goods such as wheat, maize, and soybeans. It may neither be efficient nor environmentally sound for developing countries to seek food security by becoming self-sufficient in the production of all food crops, particularly when such production involves inefficient, unsustainable methods on fragile lands.

There are numerous ways by which agricultural productivity may be increased in a sustainable way, including the use of biological fertilisers, improved pest control, soil and water conservation, and the use of improved plant varieties, produced by either traditional or biotechnological means. Of these measures, biotechnological applications, especially transgenic plant varieties and the future products of functional genomic projects, probably hold the most promise for expanding agricultural production and productivity when properly integrated into traditional systems. It can therefore be expected that local food production will need to depend on non-organic methodologies unless conventional breeding techniques overcome limitations for fertiliser demand and yield in the very near future, or organic agriculture embraces genetic modification to speed up its development.

5.3.3.1
Role of Plant Biotechnology

There is considerable debate about whether conventional plant breeding can continue to generate yield increases and provide farmers with ways of reducing input constraints. So far conventional plant breeding still leads in contributing to yield increases. Farmers were modifying plant genomes long before they knew anything about genes. For thousands of years, they sought to transfer desirable traits from one plant species to another by cross-breeding. This was how wheat was obtained from wild grasses. Modern plant breeders further developed these crops with additional desirable traits, such as disease resistance. When crossing two plants, hybrids are formed and their genetic materials will be fully mixed. After subsequent rounds of self-crossing of the progeny and selection of promising candidates, a new and stable variety may be obtained with a set of desired characteristics. It typically takes 8–12 years to produce an improved plant variety by way of conventional cross-breeding.

Now, using techniques of molecular biology, scientists can identify traits from similar species or even from unrelated organisms and, basically in a single step, transfer those desirable genes into agronomically important crops. Therefore, as an extension of traditional plant breeding, plant biotechnology or genetic modification uses genetic knowledge and scientific techniques to add specific traits to crops, such as an ability to fend off pests, survive droughts, delay ripening, or require less pesticides. Such traits specifically benefit farmers, including those in developing countries where crop losses due to weeds, pests and diseases are high and conventional tools to ward off those problems are unavailable or unaffordable. Other traits, such as disease resistances or less dependence on fertilisers also have benefits for the environment, while crops with enhanced levels of health-promoting substances or improved taste benefit the consumer.

A major advantage of GM technology is that it allows the transfer of traits between unrelated species, something that is not possible through breeding. Additionally, these plant biotechnological techniques offer strategies for crop improvement that can be applied to many different crops. A typical example includes resistance to leaf or yellow rust, a disease caused by a fungus in wheat, which has eluded science and plant breeders for a long time. Yet one grain, rice, is rustproof and it is hoped that the genetic basis for that resistance in rice can be found and transferred to other grains, like wheat. With 70% of the world's food coming from grains, such disease-free transgenic crops would be an important step towards preventing famine and producing high quality food, with additional benefits for the environment.

There are also a number of uncertainties that accompany the introduction of GM technology. There may be environmental risks involved when releasing GM organisms and there are concerns about human and animal health when new GM products are used in food or feed. Furthermore, it may turn out that farmers of the developing nations will not be able to afford the new crop varieties since GM is an expensive technology, which can only be carried out in well-equipped laboratories and needs further development and marketing by efficiently organised companies.

Also, if a limited number of successful GM crops are accepted universally, crop uniformity may reduce genetic diversity and introduce the risk of explosions of specific pests. The need to use successive novel crop varieties may prevent farmers from relying on their current practice of saving seeds for planting the next year's crop. It is feared that patenting laws work against the poor around the world and allow biotechnology companies to benefit from patenting indigenous knowledge or materials, often without consent of the original owners of the land where the valuable genetic materials were collected. Enforcing intellectual property rights over living organisms would further divert Africa's resources from where they should be directed – to feeding the poor.

Agriculture in most parts of the world is more than 5,000 years old and mass introduction of non-local high-technology crops may cause the loss of the traditional knowledge base of farmers. GM has had more to offer to the developed world than to the developing countries. So far, a restricted number of life-science companies have delivered a limited number of transgenic crops with a small number of different traits to the market place. Most of today's GM crops carry traits that are of minor importance in developing countries, or concern crops that have economic value in feeding livestock, instead of providing food for hungry people. Because companies have the task of making a profit, their primary focus is not likely to be on the needs of people that are not able to pay. Therefore, in order to serve poor farmers, plant science, breeding and genetic engineering research also needs to be directed at the staple crops they grow and which are their main source of nutrients. These include white maize, cassava, sorghum, millet and sweet potato. Improvements would also need to include adapting crop species to the adverse environments the farmers face, be applicable for small farms and not require expensive inputs farmers cannot afford.

It must be realised, however, that GM crops cannot solve the problems of improper distribution of food resources that are due to political conditions or inequalities of trade. The purpose of plant biotechnology is to enhance the ability to increase food production relative to population growth, to cope with changes in climatic conditions, to continue to outpace pests and diseases, to provide environmental improvements and to extend crops into ecologically challenging areas, such as those with acid or saline soils. It would also be sensible to use GM crops in conjunction with the practice of organic agriculture and take advantage of the best elements of two currently opposing agricultural methodologies.

Often the debate between the proponents and opponents of GM plants circles around very simplified statements. This can to some extent be attributed to its broad spectrum, involving a mixture of technological, political, economic, ethical, environmental and health issues – but this is also a reflection of a confrontation between fundamentally different ideologies and approaches. It must be realised that the transgenic crops being cultivated today will constitute only a small part of the rapidly expanding plant biotechnology portfolio with, before the end of this decade, the expected availability of many novel plants and plant products. In the future GM crops will certainly play a role in the delivery of medicines, vaccines, and improved nutrition. Thus, it would not be an exaggeration at all to conclude that new GM crops and products will display increasing consumer and environmental relevance.

In general, the proposed risks of employing GM crops is rather speculative. It is for instance impossible to prove conclusively that GM foods are 100% safe, but this is also true for food products from conventional or organic agriculture. Still, it is essential to carefully predict and test the effects of employing novel GMOs in agriculture, and to keep monitoring possible offspring afterwards. But so far, there is no evidence that GM crops are prone to cause damage to the environment or endanger human health.

Should the agricultural sector remain unchanged while every other aspect of life is changing? While not many experts suggest that transgenic crops are the single magic bullet to solve the future food problem, many scientists find it is one tool among many that can help reduce poverty and provide new innovative products. Transgenic crops can substantially contribute to improve agriculture in parallel to the needed changes at the socio-political level. Benefits and risks associated with particular GM crops can only be assessed on a case-by-case basis when employed in the field. A fair comparison to other alternatives would be the right attitude to evaluate novel crops. To do nothing, as has been the case during the European de facto moratorium on GM market introductions, is itself not without risk. An unfounded restriction of further development and employment of techniques and results from molecular biology may do more harm than good.

5.3.3.2
Role of Global Policies

More than half of the world's poorest people have little alternative but to continue to rely on agriculture for their livelihood and try to increase productivity to secure continued existence. In a global sense, the most effective strategy to ensure sufficient levels of food production would be to raise productivity in areas of subsistence farming, where an increase in food production is urgently needed and where crop yields are significantly lower than those obtained in other areas of the world. In most of these environments, crop failures are frequently due to drought and other natural catastrophes. The new tools of biotechnology, in particular genetic modification, promise the potential for creating crop varieties that are more tolerant of drought or saline soils and meet the challenges of farming.

Consumer groups claim their right to avoid GM food and see no reason for GM technology since there is sufficient food already. Will the poor have the choice to use plant biotechnology and possibly decide whether they will have food to eat or not? Will political and economic interests, for or against GM food, allow us to reach the levels of food production necessary to feed the growing world population? GM crop technology development is likely to benefit the poor only if proper technology is developed in a responsible manner and put into the right hands. The potential of GM crops should therefore not imply rejecting organic agriculture, nor disregard the value of indigenous knowledge.

Consumer resistance to GM crops in Europe, the subsequent moratorium on GMOs and dramatically reduced numbers of GM crop field trials have had an indirect effect on exporting countries in the developing world. Some countries could benefit in the short term by staying GM-free and retain their access to European markets. The needs of Europe, having surplus food, are different from Africa where we witness hunger and starvation. The priority of Africa is to feed its people with safe foods and to sustain agricultural production as well as the environment. Over the long run, when markets have harmonised and GM products find more acceptance, it may become essential for Africa and economically necessary for its farmers to reap the benefits that GM crop innovations provide.

Investing in more research and development of improved crops therefore is a promising avenue for accelerating progress and reducing global poverty. Public institutions can help in doing quality research that will benefit poor farmers, alleviate food shortages and reduce poverty. However, an ultimate failure to end hunger in developing countries will arise not from technological limitations but from political and/or economic decisions and the disinterest of governments and corporations. There is no question that the primary responsibility for dealing with poverty actually belongs to developing countries themselves and that the establishment of a fair trade regime is in the hands of the developed countries.

Self-sufficiency of developing countries can only be reached if there are no restrictions on export markets for agricultural products under equal competitive conditions. Currently, the total subsidies to agriculture in developed countries are six times the amount of development aid that is given to poor countries. Developed countries therefore produce surplus food and readily supply this in the form of aid, as a result of which the developing countries do not get adequate pressure to adopt effective food policies themselves. The anomaly of this agriculture policy is that rich countries tend to subsidise a declining agricultural sector, which may contribute less than 5% to GDP, while poor countries, where agriculture is the dominant sector, are limited in possibilities to expand economically.

Globalisation is not an automatic blessing and will certainly not eradicate poverty on its own. While international trade and investment have increased vastly over the last decade, this process has not really been global enough. Two billion people live in countries where trade is diminishing in relation to national income, economic growth has stagnated, and poverty is on the rise. Most people in Africa, Latin America, the Middle East, and Central Asia are poorer today than they were 10 years ago. In many of these countries the systematic restriction of economy and political freedom are caused by any of the following: an over-regulated economy, corruption, repression or war.

In theory, sound macroeconomics and market liberalisation, as proposed by the World Bank and the IMF, together with freedom and democracy seem to form the complex recipe for a globalised capitalist economy to provide opportunities for developing countries to improve integration with the rest of the world. It is unfortunate that, in practice, these arrangements tend to be bent to favour specific market niches or political goals and, of course, by the most powerful parties. Nevertheless, people in Asia have enjoyed a big improvement in living standards, especially since the 1980s, when China and India began to adopt market reforms. Much of the improvement in Chinese economic performance can indeed be attributed to liberalisation since its international trade expanded when the economy became market-oriented and less regulated.

While globalisation may have diminished a rising inequality between participating nations, it requires a significant policy change to properly exploit the strategy. Nations that did not adopt these changes, or were too isolated to do so, were left behind. In theory, globalisation should give small farmers access to lucrative world markets, but it is equally possible that in a number of countries they lose much of the urban markets of their major cities to imported goods from other continents.

5.3.3.3
Role of Companies

The top ten global seed companies control almost one-third of the $25 billion commercial seed trade. The concentration of patents and plant variety owner-ships make it difficult for small start-up companies or public sector researchers to compete or gain access to new agricultural technologies. The multiplicity of patents embedded in a GM product can make the product expensive. It is also possible that the holder of a single piece of intellectual property right (IPR) decides to block the commercialisation of a rival product. The costs associated with transacting IPRs can be quite substantial and may give private companies considerable power because of their strategy of building up defensive patent portfolios. Critics claim therefore that poor countries have no chance to benefit from plant biotechnology or that poor farmers will only be exploited by multi-nationals. The current patent system, in addition, does not offer sufficient pro-tection for the genetic material of local varieties and local community-based innovations.

It is, however, not very likely that patents in biotechnology will be used to seriously frustrate indigenous plant breeding. There is a misplaced concern that patents and other forms of intellectual property constrain the freedom to operate of plant science researchers in developing countries. For most IPRs in plant sciences that are held by life science companies, no protection has been sought in the majority of such countries. Crop breeders are free to produce any crop as long as the inputs and processes used and the crop varieties grown are not protected under local intellectual property laws. The restriction is that those crops cannot be legally exported to countries where they fall under intellectual property protection. Intellectual property rights assigned to the key enabling technologies, for instance those used to transform crops, are primarily relevant to rich-country jurisdictions. For many of the crops that matter in poor countries and of which little is exported (e.g. cassava, coconut, groundnuts, beans, sorghum, lentils) much of the needed technology is not covered by intellectual property rights locally and can be employed without much restriction for local use.

For researchers and farmers in less-developed countries it is therefore much more important to gain access to technologies in plant biotechnology than to worry about possible, future, IPR problems in international trade. Failure to attract investments in domestic expertise, needed to evaluate, access, and regulate the new technologies, is currently a far greater constraint. Poor people rely for sustenance on crops that are largely beyond the focus of the private research sector and that have modest future commercial prospects. In addition, poor producers often face production problems different from those of commercial farmers in wealthier countries. Companies are producing and marketing new crop varieties for a profit and therefore have shown little interest in developing the crops most often grown by subsistence farmers (such as millet, sorghum and rice) or design others that would survive the harsh weather and soil conditions in needy countries.

As in the case of medicine and pharmacology, major advances have often come from large corporations with an eye firmly on a return on investment and shareholder dividends. In the past few years a number of large life science companies nevertheless have shared information or donated intellectual property for use in certain non-commercial crops. Monsanto shared its draft rice genome sequence data with public researchers involved in the International Rice Genome Sequencing Project. Syngenta, among others, donated IPRs relevant to vitamin A rice to allow free licensing to farmers so long as they do not export such products or profit excessively from it. These well-publicised donations left a strong impression that a large number of crucial patent rights were transferred in favour of the poor in developing countries. But, in fact, in some major rice-consuming countries, there are no relevant valid patents and in most countries there are very few.

In any case, companies have seen it in their own interest to develop partnerships with the public sector that will help bring benefits to developing countries. An initiative by the Rockefeller Foundation, known as the African Agricultural Technology Foundation (AATF) will function as a broker of public–private partnerships, and will act as a focal point for materials and information on technologies. Monsanto, DuPont, Syngenta and Dow Agrosciences have announced that they will donate research tools, seed varieties, patent rights and laboratory techniques free to African scientists through the AATF being established in Kenya. Among the aims of the organisation are:

1. Involvement of farmers in the design of new crop varieties
2. Find effective combined use of organic with inorganic fertilisers
3. Make African products more competitive in global markets
4. Implement the new technologies of plant biotechnology

5.3.4
Regulation

The global debate about the acceptability of GM foods came to international attention when a number of African countries that, although suffering famine, refused US food aid in 2002 because it included GM maize. The USA food aid program uses commodity maize, which typically includes the GM maize that is widely grown in the USA. This situation highlights the global ambivalence over a technology that some have touted as a tool to end world hunger and others have criticised as providing no benefits to consumers while posing significant risks to the environment.

The reluctance of some African governments to accept GM maize is not representative of the attitude of developing countries towards biotechnology. Ethiopia is wary of GM while Kenya, its neighbour to the south, is actively pursuing the technology as a means to increase food security. South Africa and Argentina have embraced GM agriculture. China is actively pursuing its own GM research programme and India has just approved GM cotton. In 2002 nearly 10% of the global acreage of transgenic crops was planted in the developing countries. This fact would negate the argument that plant biotechnology is a tool of the industrialised countries for the exploitation of the developing world, although it must be realised that most GM crop seeds currently originate from western breeders.

5.3.4.1
Towards Consensus

How could world wide consensus on the regulation of agricultural products be achieved? The World Trade Organization's (WTO) legal framework regarding trade in GM products only includes the Sanitary and Phytosanitary (SPS) Agreement and the Agreement on Technical Barriers to Trade. The SPS Agreement says that WTO members have "... the right to take sanitary and phytosanitary measures only to the extent necessary to protect human, animal or plant life or health, and based on scientific principles...". The SPS Agreement stipulates that food safety regulations be scientifically justifiable and requires WTO member governments that violate the SPS Agreement to modify or withdraw their non-compliant food safety regulations. In 1995, the SPS agreement conferred on the Codex Alimentarius commission the responsibility for defining international food safety standards that would be recognised by the WTO.

In the absence of agreed-upon international standards, some countries invoke the precautionary principle that allows them to rule provisionally where relevant scientific evidence is lacking, although they are supposed to do the necessary research within a reasonable period of time. Other countries argue that the precautionary principle is being abused in order to protect less efficient domestic producers from foreign competition. Europe has adopted the precautionary principle as a means for determining whether GM crops can be planted or whether a food product is safe enough to eat. The principle states that when an activity raises threats of human health or harm to the environment, precautionary measures should be taken even if cause and effect have not yet been scientifically established.

The WTO Agreement on Technical Barriers to Trade (TBT) aims to ensure that WTO members do not use domestic regulations, standards, testing, and certification procedures to create unnecessary obstacles to trade. The TBT rules encourage countries to use international standards as a basis for national laws and these include provisions for labelling. Since the WTO SPS or TBT rules do not interpret the precautionary actions taken by Europe and some other countries to restrict or label GM food imports, there is room for conflicting interpretations. More than 30 countries currently require labelling of GM foods. Europe has one of the strictest labelling standards, requiring labels if more than 0.9% of the ingredients in foods are GM. It is this labelling standard that motivated the African countries to refuse the US food aid that contained GM maize. There was concern that farmers would plant the maize instead of using it as food, resulting in cross-pollination of native maize seeds. If that should happen the European standards would no longer be met and the countries would no longer be able to export surplus maize in the future. The US government has brought a dispute against the EU's refusal to accept GM crops and products to the WTO and this procedure may clarify the grey areas in WTO rules one way or the other.

The United Nations-sponsored Cartagena Protocol on Biosafety (CPB), which was adopted in January 2000 in Montreal, regulates the trade of living GM organisms (LMOs) and requires member nations to establish strict controls over many aspects of research and development on GM crops, animals, and micro-organisms. Article 1 of the CPB refers to the precautionary approach and to "…ensure an adequate level of protection in the field of safe transfer, handling, and use of living modified organisms resulting from modern biotechnology…" Article 2 emphasises the right of states to enforce more stringent protection goals for the "…conservation and sustainable use of biological diversity than called for in this Protocol". For exporters, the protocol is potentially troublesome because it allows importers to refuse LMOs on a precautionary basis, without a scientific demonstration of risk. The United States, which is not party to the Cartagena pact, treats GM grains no differently from conventional crops and says they pose no threat.

Close to 200 countries and regions have signed the Cartagena protocol, and as soon as 50 ratifications are completed, it will come into force. The CPB will be the first legally binding international agreement that will make use of the precautionary principle as a policy instrument for the conservation and sustainable use of biological diversity and the protection of human health, if relevant to the trans-boundary movement of GMOs. Risk assessment remains the primary tool under the CPB for competent authorities to "…identify and evaluate the potential adverse effects of living modified organisms on the conservation and sustainable use of biological diversity in the likely potential receiving environment, taking also into account risks to human health". The precautionary principle is concerned with possible unacceptable risks and threats and calls for risk assessment to be carried out. Both measures are compatible if risk assessment can provide some indication as to the proper degree of precaution that is warranted and if precaution can be used to audit and refine risk-assessment assumptions.

While many countries have biosafety regulations or laws on labelling, the majority of them do not. Building strong biosafety capacity with clear guidelines, systematic assessment, public consultation and product labelling can help countries to make informed decisions. Many developing countries do not have the multiple- and inter-disciplinary personnel needed to carry out risk analyses and risk management within a methodological framework, as stipulated by modern regulations. Assistance to such countries that attempt to establish a national biosafety capacity therefore deserves priority support. It remains to be seen whether the world community will be able to accept modern plant biotechnology in some regulated form or whether reservations, prevalent at present in Europe, will continue to slow down the development and assessment of its potentials.

5.3.4.2
Intellectual Property Rights: Patents and Plant Variety Protection

If information, new ideas or inventions are published without proper protection, these will then enter public domain, meaning that anyone can use the information without asking for permission. Developments in technologies that are based on public knowledge will not be supported by large investments because the possibility of imitation creates a discouraging commercial future. For this reason, a number of different mechanisms have been established for the protection of, for instance, information (copyright), inventions (patents) or brand names (trademark).

When corporations or, increasingly, also public agencies develop new crop varieties, they often seek intellectual property right (IPR) protection on these innovations. Inventors and researchers seeking these rights must then disclose the new knowledge they have obtained, which will stimulate further rounds of innovation and technological advances. Patents and other IPRs are awarded by national governments or common trading blocks. To obtain patent protection in more than just one country, innovators must separately apply and gain such rights in additional nations.

Similar restrictions apply more or less for protection through the plant variety protection laws (PVP, or plant breeders' rights), which are generally based on the UPOV Convention. To be granted protection, a variety must fulfil criteria governing novelty, distinctiveness, uniformity and stability while allowing farmers to save seeds for their own use. PVP also defines "breeders exemption", which allows any plant breeder to use the protected variety as a basis to develop a new one without previous consent of the owner of the original protected variety.

IPR regimes tend to be weak in developing countries, but the WTO agreement on Trade-Related Aspects of Intellectual Property Rights (TRIPS) prepares for increased global harmonisation. This international agreement is binding on all WTO members and sets minimum standards for the implementation of IPRs at the national level. While some argue that the agreement favours the biotechnology industry in developed countries, an important provision is that plants and animals may be excluded from patentability by national governments. Although WTO members are therefore not required to allow patents on living organisms, a minimum requirement will be to allow protection of plant varieties through the PVP system.

The impacts of IPRs are more substantial for modern biotechnologies and products with multiple patents, such as GM plants, than for new plant varieties obtained through breeding or for products that are derived from other biotechnologies, such as micropropagation or tissue culture. GM technology may increase the dominance of corporations in the area of food production, particularly if strict intellectual property regimes are permitted globally. Plant breeders' rights, rather than patents, should be favoured in less-developed countries in order to protect the interests of poor farmers. Poor countries should therefore be wary of any provisions in trade deals that try to impose stronger intellectual-property standards than TRIPS requires. Rich countries should accept that considerations of how intellectual-property rights affect poor countries are not just a concern of overseas-aid agencies, but also play a part in broader trade and economic relations.

5.3.5
Regional Example: China

China was the first country to begin growing a GM crop commercially with a virus-resistant tobacco in 1988, following a public–commercial biotech collaboration between the Beijing University and the government tobacco companies. Since then, China has tested over a hundred GM crop varieties and is making major investments through its institutions such as the Chinese Academy of Agricultural Sciences and the Chinese Academy of Sciences. China has dramatically accelerated innovation and commercialisation in plant biotech research. Nearly all scientists are employed in the public sector, therefore it is easier to undertake research on crop technologies. By making good use of the complementary benefits of biotechnology and traditional crop breeding, the Chinese public research programme is making progress on many fronts, from sequencing the rice genome to developing hybrids for wheat, soybean, rapeseed and other crops. The broad spectrum of traits that is addressed by Chinese researchers include yield increases, disease resistances, abiotic stress responses, enhancement of the nutritional value of crops and reduction of the use of inputs that are causing environmental and health concerns. China's biggest genetic effort focuses on rice, the worlds most consumed grain. The China National Rice Research Institute in Hangzhou is performing leading research on engineered rice varieties with better yields, nutrition and taste, and improved drought and insect resistance. Although only an eighth of China's land is suitable for cultivation, it is among the worlds largest producers of rice, barley, sorghum, potato, peanuts, tea, fruits and vegetables. Analysts therefore predict that further progress in biotechnology could turn China into a global competitor, having a major commercial influence in agro-biotechnology. In non-food use of GM crops, insect resistant cotton tops the list of field releases.

Amidst all this activity, Chinese regulators have, since 2000, been reluctant to approve the commercial planting of transgenic food crops, although candidate crops continue to move into field trials. Before this intervention there was no government control over when GM crops graduated from the research to the commercial arena, nor were there labelling requirements when GM foods were sold in the market. China's new GMO regulations of May 2001 require labelling and safety certification of all GM products, and also cover GMO research, testing, production, processing, import/export activities and liability when handling GMOs. While the government claims to be worried about whether Europe will accept exports of transgenic food, it is also assumed that the Chinese government is exploiting the biosafety issue to frustrate the commercial ambitions of Western agribiotech firms, because it realises that its own research programme needs more time to catch up.

5.3.6
Product Example: Golden Rice

An observed correlation between progressive and serious eye damage in children and increased child mortality rates in Indonesia prompted researchers to supply vitamin A capsules. The startling result of the significant decline in child mortality rates, common in developing countries, was confirmed with similar experiments in Africa and Asia. While vitamin A is primarily essential for the general health of the eye, it has many secondary roles, such as maintenance of the immune system. Vitamin A deficiency is prevalent among the poor in Asia, because their diets are very much dependent on rice, which does not contain beta-carotene, a vitamin A precursor. While individuals may be able to obtain sufficient vitamin A by consuming larger amounts of dark-green leafy vegetables and fruits, vitamin A-deficient populations live in marginalised areas and have poor diets (see also Sect. 5.1.2.5.3).

It is now widely accepted by the international nutrition community that the distribution of vitamin A is a priority for government intervention. Nutritionists in developing countries have been able to demonstrate that many children and young women suffer more from a lack of essential vitamins and minerals in their diets than from a lack of calories. Poor dietary quality has therefore been dubbed "hidden hunger" and stimulated further research related to other micronutrient deficiencies, in particular iron and iodine deficiencies.

Golden Rice was developed to provide a new, alternative intervention to combat vitamin A deficiency. Peter Beyer was studying the regulation of the terpenoid pathway in daffodil and isolated the relevant genes of the beta-carotene synthetic pathway. Researchers in the laboratory of Ingo Protrykus at the Swiss Federal Institute of Technology succeeded, against all odds, in the introduction of the genes and the co-ordinated function of the four identified enzymes in rice. The endosperm of the grains of the GM rice was coloured yellow, and was shown to contain beta-carotene and other carotenoids, which the human body uses to convert into vitamin A. The dream of the inventors is that golden rice, once adapted to fields in developing countries, would be able to deliver the vitamin A precursors in a preferred food product, avoiding the complicated institutional infrastructure of vitamin A supplementation and fortification.

The inventors, however, realised that further development and marketing of the GM rice would be impossible without the help of a commercial partner. They signed an agreement with Syngenta to provide the necessary technology and support, for free, to the needy people in developing countries. In return, the company would able to explore commercial opportunities for the sale of golden rice into the expanding market for healthy foods in more developed countries. Syngenta would also be providing regulatory, advisory and research expertise to assist in making golden rice readily available among developing nations.

There is a range of challenges and further developments that will need to be addressed in the next few years. Since the major rice varieties for consumption are of the indica type and golden rice is a japonicum, breeding efforts are underway to obtain varieties that will deliver similar or better yields. Secondly, the amount of carotenes delivered in the grains needs to be raised in order for golden rice to have sufficient impact. Since wheat consumption is rising throughout Asia, it may also be necessary to fortify wheat varieties. Then there are still important questions to be answered in relation to food safety, consumer acceptability, and biodigestibility. Several governments have indicated their financial support for making future golden rice varieties available to farmers, but the implementation of these plans depends on reaching global agreements on GM technology, favourable regulatory regimes and consumer acceptance.

It appears that golden rice has the potential to be a low-cost, wide-coverage intervention to mitigate the effects of vitamin A deficiency. While it may significantly contribute in this battle under certain scenarios, it is unlikely to meet all requirements and to form a stand-alone strategy. Therefore, further research is needed as well as exploring other possibilities such as expressing genes to raise iron and vitamin E levels or the development of a high-quality protein rice with enhanced levels of selected amino acids. Ingo Protrykus has expressed the opinion that the golden rice project has been an excellent example of a public–private partnership that has real benefits for the poor and that public sentiment can only be improved if more such projects were developed in public institutions, using public funding and addressing an urgent need that cannot be solved with traditional techniques.

5.3.7
Conclusions

Biotechnology offers tools that allow plant breeders to generate superior plant varieties, or select these much faster than they could when restricted to using conventional plant breeding techniques. While consumers in developed countries have the right and luxury to debate the pros and cons of products from GM techniques, it would be wrong to slow down basic research to study whether such technologies are safe, sustainable, and suitable for developing countries. The debate whether application of GM technology will help to feed the earth's growing population currently rests in the hands of the experts in developed countries. Most of the people that will need to be fed, however, live in developing nations. It is ironic to note that modern agricultural practices and technologies have resulted in the most abundant, healthiest, and cheapest supply of food in the history of the human race. Therefore, any government should be allowed the right to make their own decisions on biotechnology, which they cannot do if access to such technology is denied to them or trade issues prevent such developments.

Unlike the techniques of the green revolution, GM technology was largely developed by private companies. In the minds of some people the mingling of development and profit makes the technology suspect. However, the profit motive is a strong incentive to produce a healthy product. At the same time, there is no sustained and coherent effort to make this technology accessible and usable by people who cannot afford such technologies and products. Moreover, if only large multinational companies gain control over the food chain then the production technology may become driven by the economic logic of delivering higher returns and not by concerns over the environmental health or sustainability of the planet.

Food quality and food security should be an essential human right. Therefore plant biotechnology must be adequately supervised, its technologies accessible to all involved in agriculture, and inventions as well as indigenous genetic resources should be protected. In view of these points it must be concluded that there is still a great deal to be accomplished by politicians as well as scientists in public research institutes.

5.3.8
Organisations and Sources

Below is a far-from-complete list of national and international organisations involved in different aspects of technology transfer of agro-biotechnology between
developed and developing countries. While the number of research co-operations
and other initiatives is much more extensive, such a list is deceptive. The total
investment in agricultural research in many countries is in crisis and funding
levels are insufficient to generate a steady flow of technology.

5.3.9
Not-for-Profit Organisations

- AfricaBio
  The International Service for the Acquisition of Agri-biotech Applications, and
  other stakeholder organisations. http://www.africabio.com
- APEC
  The Asia-Pacific Economic Cooperation has a task force that deals with issues
  related to biotechnology, such as biodiversity protection and the safety of
  GM food.
- BIO_EARN
  The East African Regional Programme and Research Network for Biotech-
  nology, Biosafety and Biotechnology Policy Development is co-ordinated by
  the Stockholm Environment Institute (SEI) to build national capacity and
  competence in biotechnology, biosafety and biotechnology policy involving
  more than 70 researchers and more than 100 policy makers in the region.
  Selected institutes in Ethiopia, Kenya, Tanzania and Uganda receive support
  through a regional network. www.bio-earn.org
- CAMBIA
  Centre for the Application of Molecular Biology to International Agriculture
  is a not-for-profit research institute in Canberra, Australia, which was set up
  in 1991 to develop and package the novelty generation and selection tools that
  biotechnology is making possible so that farmers and local researchers can use
  them. www.cambia.org
- CAST
  The US Council for Agricultural Science and Technology is an international consortium of 38 scientific and professional societies. Its mission
  is to identify food and fiber, environmental, and other agricultural issues
  and to interpret related scientific research information for legislators,
  regulators and the media for use in public policy decision making. www.cast-
  science.org
- CGIAR
  The Consultative Group on International Agricultural Research is an alliance of
  24 developing and 22 industrialised countries, four private foundations, and
  16 international agricultural research centres known as The Future Harvest
  Centers, among which IRRI, CIMMYT, IITA and IPGRI. The CGIAR is best
  known for starting the green revolution of rice and wheat in Asia. www.cgiar.org

- CIAT
  The Centro Internacional de Agricultura Tropical conducts international research on beans, cassava, and forages has a global reach, while that on rice and tropical fruits targets Latin America and the Caribbean. www.ciat.cgiar.org
- CIMMYT
  Using traditional breeding methods, scientists from the International Maize and Wheat Improvement Association, a Future Harvest Center, and their South African counterparts set out to develop corn that could increase the food and income of poor southern African farmers who grow food in drought-prone, nutrient-depleted soils. www.cimmyt.org
- CIRAD
  Centre de Coopération Internationale en Recherche Agronomique pour le Développement is a French scientific organisation, with researchers posted in 50 countries, specialising in agricultural research for the tropics and subtropics of the world, focusing on crops like rice, cotton, coffee and rubber. http://www.cirad.fr/en/index.php
- FBAE
  The Foundation for Biotechnology Awareness and Education, based in Bangalore, India, supports sustainable development through biotechnology by promoting biotechnology awareness and education. www.fbae.org
- IITA
  The International Institute of Tropical Agriculture, based in Nigeria, is one of the CGIAR research centres. By linking advanced research institutions around the world to six Benchmark Areas in countries in Africa, it will be able to make use of the shared benefits of biotechnology. www.iita.org
- IPGRI
  The International Plant Genetic Resources Institute, a Future Harvest Center, is based in Rome www.ipgri.cgiar.org/One of its programs is the International Network for the Improvement of Banana and Plantain (INIBAP). www.inibap.org
- IRRI
  The International Rice Research Institute, a Future Harvest Center, aims to ensure that rice farmers and consumers get the best deal and the best options offered by science and the private sector, while helping companies find ways to get the returns they need to support the further development of their activities and the rice industry. www.irri.org/default.asp
- ISAAA
  The International Service for the Acquisition of Agri-biotech Applications has established the Global Knowledge Center on Crop Biotechnology to share crop biotechnology information with as many people as possible. When fully operational, the Knowledge Center will consist of at least 20 country nodes, and will facilitate information sharing between and across these countries. www.isaaa.org
- ISAS
  The International Society of African Scientists is a non-profit organisation founded in 1982 to promote the advancement of science and technology among

people of African descent, and to solve technical and development problems facing Africa and the Caribbean. www.dca.net/isas/
- ITDG
The Intermediate Technology Group is an international non-governmental organisation specialising in technology transfer for Practical Answers to Poverty. ITDG works in Latin America, East Africa, Southern Africa and South Asia, with particular concentration on Peru, Kenya, Sudan, Zimbabwe, Sri Lanka, Bangladesh and Nepal.http://www.itdg.org
- MARDI
The Malaysian Agricultural Research and Development Institute was established in 1969 to produce and promote appropriate and efficient technologies that could improve production and income from agriculture. www.mardi.my
- REDBIO
The Technical Cooperation Network on Plant Biotechnology focuses on bringing biotechnology to the service of sustainable development of the Latin America and Caribbean forestry and agricultural sectors. www.redbio.org
- TWN
The mission of the Third World Network, established in 1984, is to conduct research on economic, social and environmental issues pertaining to the South, to publish books and magazines, to organise and participate in seminars and to provide a platform representing broadly southern interests and perspectives at international fora such as the UN conferences and processes. www.twnside.org.sg
- UNFPA
The United Nations Population Fund helps developing countries collect and analyse population data and to integrate population and development strategies into national, regional and global planning. www.unfpa.org

5.3.10
Information Sources

OECD (2003) Accessing agricultural biotechnology in emerging economies. OECD, Paris. http://www.oecd.org/pdf/M00041000/M00041637.pdf
ICSU (2003) New genetics, food and agriculture: genetic discoveries – societal dilemmas. ICSU, Paris. http://cbac-cccb.ca/epic/internet/incbac-cccb.nsf/en/ah00335e.html
ESTO/JRC (2003) Review of GMOs under research and development and in the pipeline in Europe. JRC, Brussels. http://www.jrc.es/gmoreview.pdf
Nuffield Council on Ethics (2003) The use of genetically modified crops in developing countries. Draft report, October 2003. Nuffield Council on Ethics, London. http://www.nuffield-bioethics.org/filelibrary/pdf/gm_draft_paper.pdf

6 Guide to Terms

6.1
Guide to Terms

- Bioethics
  Examines the appropriateness of choices and actions taken in specific areas of biology
- Biotechnology
  Application of biology, including the field of genetic engineering, to our everyday lives
- Coat protein mediated cross protection
  Engineered viral coat protein gene is introduced into host plant to protect crops against the challenge of infection
- Consumers
  People who use products
- Criteria of risk
  Frequency and scope of a harmful event occurring
- Cross protection
  Indicates that the protective action is reciprocal
- DNA
  Deoxyribonucleic acid (sometimes called nucleic acid). A biological polymer that contains and transmits, through replication and transcription, the genetic information of the organism. It is composed of nucleotide units called bases (A adenine, T thymine, Q guanosine, C cytosine). It is the specific order of these bases that can code instructions for what the organism will be like
- Ethics
  The use of a rational approach to examine and analyse moral concepts, questions and resulting choices and actions in a specific area or situation. Whereas what is considered as moral behaviour may sometimes differ from region to region, rules of ethical behaviour should be universal. Thus what may be moral may not be entirely ethical, however, what is ethical always contains a subset of moral concepts. In effect, ethics helps to define and incorporate the universal core of moral behaviour. For a medical doctor ethical rules mean, for example, to be helpful and do no harm, to respect a patient as a person and to be non-discriminatory
- Facts
  Observations that can be measured, tested and verified

– Gene manipulation
Recombinant gene technology or molecular cloning
– Genetic engineering
Formation of new combinations of hereditary material by processes that do
not occur in nature. The technology is sometimes called modern biotechnol-
ogy, gene technology, gene cloning or recombinant DNA technology and often
refers to genetically modifying living organisms
– Genetic information
Sum total of hereditary information that is needed for a species to survive
from generation to generation
– Genetically modified organisms (GMOs), including GM plants (GMPs) and
GM micro-organisms (GMMs)
Living organisms in which the genetic material (DNA) has been altered in a
way that does not occur naturally by mating or natural recombination. The
use of genetic engineering allows selected individual genes to be transferred
from one organism into another, even between non-related species
– Genome
Complete set of genetic instructions of an organism. The instructions exist as
specific sequences of DNA or RNA
– Harm
Damage
– Hazard
Source of danger identified on the basis of some intrinsic properties or prob-
ability of occurance
– Host
Organism harbouring and supporting growth of another organism. The
relationship could be parasitic (benefiting only of the two organisms) or sym-
biotic (benefiting both organisms)
– Morality
Decision-making process based on attitudes that help to distinguish correct
and incorrect choices and actions, thus in the process defining the character
of the individual, group or a society. Morality is influenced by religion, regional
societal values and beliefs
– Nucleocapside
Nucleic acid component of the virus particle together with its surrounding
protein shell (capsid)
– Plantibody
Genetically engineered plant producing antibody against infecting viruses
– Precaution
Prudent foresight; actions taken to ensure good results
– Probability
Likelihood of an event taking place
– Proteins
Large molecules composed of amino acids. There are 20 amino acids that can
form proteins. Any combination of amino acids can be used for the creation
of proteins. This depends on a complex process that starts with the decision
about which genetic information of an organism is to be transcribed. Proteins

are essential for the existence of living organisms. For example, all enzymes that enable cellular processes to proceed, are proteins
- Public
Concerning the people as a whole
- Recombination
Exchange of genetic material (DNA or RNA) between two individual organisms, resulting in a change in genetic makeup and properties. The exchange is heritable and permanent
- Replication
Copying of the genetic material
- Risk
Exposure to danger, sometimes also defined as the probability of harm. The risk can be voluntary (accepting and knowing the dangers involved), nonvoluntary (not knowing the dangers) and involuntary (forced into a dangerous situation without consent). Describes the magnitude of harm caused by a hazard and the frequency with which that hazard occurs
- RNA
Ribonucleic acid (sometimes called nucleic acid). A biological polymer that is usually involved in transcribing DNA information that can lead to the formation of proteins. It is composed of the same nucleotide units as DNA except for thymine that is replaced by uracil, U. In some organisms, such as viruses, RNA performs similar functions to DNA—containing and transmitting the genetic information of the organism
- Substantial equivalence
Indication that the composition, nutritional value, or intended use of GM food has not been altered compared to the non-GM counterparts
- Symptom
Visible or otherwise detectable phenotype abnormality arising from disease
- Synergism
Association of two or more viruses acting at one time and affecting a change which one alone is not able to make
- Threat
Indications of something undesirable likely to happen
- Transcription
Transfer of genetic information, usually from DNA onto RNA, the information on the latter to be translated into amino acids (see proteins)
- Transencapsidation/heteroencapsidation
Enclosure of the genomic RNA of a virus within a complete protein capsid of a second virus (it can be full or partial)
- Transgene
Gene which has been transferred into another organisms
- Transgenic plants
Plants containing artificially transferred pieces of DNA from other living organisms by means of genetic engineering
- Uncertainty
Reduced confidence in estimating the likelihood of an event taking place. Uncertainty can be of quantitative or qualitative nature

- Virion
  Result of encapsulation of the virus genetic material by a protein coat
- Virus
  Infectious sub-microscopic and filterable non-cellular agent that multiplies
  only in living cells and often causes diseases (a latin word meaning poison)
- Virus satellite
  Small RNA which multiplies only in the presence of a specific virus
- Virus strain
  Group of similar virus isolates, that are serologically or immunologically
  related

6.1.1
Information Sources

Coombs J (1986) Macmillan dictionary of biotechnology. Macmillan, New York
European Federation of Biotechnology (1997) What's biotechnology? European Federation
    of Biotechnology, Briefing paper 6. EFB
Hodson A (1992) Essential genetics. Bloomsbury, London

Subject Index

Produktion: abcdruck GmbH, Heidelberg